The Craft of Collaborativ

Unlike books that focus solely on methods, *The Craft of Collaborative Planning* provides a detailed guide to designing and managing all aspects of the collaborative process, advocating for making collaborative work the norm.

Beginning with a discussion of the political and legal context of collaborative practice in UK land use planning systems, *The Craft of Collaborative Planning* tracks a path through the challenging task of process design and working with various groups and individuals. Taking into account the great need for coherent organisational approaches, Bishop outlines evaluation and learning from the collaborative process for the future.

Jeff Bishop brings to his writing an exemplary career focused on bringing various parties together to generate creative and widely supported plans and projects. With its focused discussion of UK engagement practices and detailed outline for making a better collaborative process, *The Craft of Collaborative Planning* is an essential read for practitioners and decision-makers seeking to bring communities together with creative solutions to spatial planning, design and development.

Jeff Bishop's career in planning, design and development has focused on bringing all parties, especially the public, together to generate creative and widely supported plans and projects. Though working mainly in the UK, where his projects are often quoted as exemplars of good practice, he has also worked in other countries. Jeff is an Associate Director of Place Studio Limited.

"Jeff Bishop is not only a master craftsperson of Collaborative Planning but is also very skilled in helping others learn the ropes, in person and through his writing. I know of no other text which treats this essential *art* in such a thorough and accessible manner. It is a *must* for all those interested in the shaping tomorrow's cities."

Ray Lorenzo, City Planner, Umbra Institute, Italy

"A marvellous book by a leading pioneer. What you get is honest, straightforward, neatly illustrated, practical guidance based on the wisdom and experience of someone who has been living and breathing collaborative planning for several decades. Getting all parties to work together on planning is essential for tackling the challenges facing communities all over the world. Every aspiring practitioner will want to read this."

Nick Wates, author of The Community Planning Handbook *and* The Community Planning Event Manual, *publisher of Communityplanning.net*

The Craft of Collaborative Planning

People working together to shape creative and sustainable places

Jeff Bishop

NEW YORK AND LONDON

First published 2015
by Routledge
711 Third Avenue, New York, NY 10017

and by Routledge
2 Park Square, Milton Park, Abingdon, Oxon OX14 4RN

Routledge is an imprint of the Taylor & Francis Group, an informa business

© 2015 Taylor & Francis

The right of Jeff Bishop to be identified as author of this work has been asserted by him in accordance with sections 77 and 78 of the Copyright, Designs and Patents Act 1988.

All rights reserved. No part of this book may be reprinted or reproduced or utilised in any form or by any electronic, mechanical, or other means, now known or hereafter invented, including photocopying and recording, or in any information storage or retrieval system, without permission in writing from the publishers.

Trademark notice: Product or corporate names may be trademarks or registered trademarks, and are used only for identification and explanation without intent to infringe.

Library of Congress Cataloging in Publication Control Number: 2014045733

ISBN: 9781138840409 (hbk)
ISBN: 9781138840416 (pbk)
ISBN: 9781315732848 (ebk)

Typeset in Goudy
by Saxon Graphics Ltd, Derby DE21 4SZ

Contents

List of figures		vii
List of tables		ix
Acknowledgements		xiii
1	Introduction	1
2	Setting the Scene	9
3	Getting Ready	42
4	Design to Deliver	67
5	Delivering into Detail	88
6	Working with People	121
7	Evaluating and Reporting	152
8	Making it Mainstream	173
9	Conclusions … or an Engagement Utopia?	199
Appendix 1	Collaborative Working	212
Appendix 2	The Legal and Quasi-Legal Context	217
Appendix 3	Resourcing Engagement	228
Appendix 4	Becoming a Practitioner	231
Appendix 5	Further Reading	233
Index		237

List of Figures

Figure 1	Arnstein's original ladder	12
Figure 2	Another way of seeing Arnstein's ladder	13
Figure 3	Positions, interests and needs	20
Figure 4	Compromise and consensus	21
Figure 5	Decide–Announce–Defend	22
Figure 6	Engage–Deliberate–Decide	22
Figure 7	How many might engage?	23
Figure 8	How immediate is any issue?	24
Figure 9	Inclusiveness	25
Figure 10	Basic analysis of stakeholders	76
Figure 11	A completed stakeholder analysis	77
Figure 12	Approaches derived from analysis	78
Figure 13	A blank process plan	79
Figure 14	The Waterton process plan	80
Figure 15	The three key stages	82
Figure 16	An example issues 'map'	104
Figure 17	An example mindmap	105
Figure 18	A priority-setting grid	109
Figure 19	Strategy grid cards	112
Figure 20	Blank strategy grid	113
Figure 21	Completed strategy grid	115
Figure 22	Example issues sheet	118
Figure 23	Scoring scales	119
Figure 24	Chairperson, arbitrator, mediator, facilitator	125
Figure 25	Cabaret style room arrangement	128
Figure 26	Evaluation scales	157
Figure 27	Infrastructure for engagement	177
Figure 28	The model as experienced	181

List of Tables

Table 1	A preferred levels framework	15
Table 2	Principles of engagement summarised	19
Table 3	Examples of DAD and EDD	35
Table 4	Action planning grid	144

To my wife Pat whose name very appropriately includes the first three letters of the word "patience"!

Acknowledgements

This book is built from over 40 years of experience, privileged by the remarkable range of settings in which I have been able to work, by the sharing (some might say borrowing) of ideas and challenges with my peers and by the contributions through valuable feedback (all informative if not always positive) from the now thousands of participants at many, many events.

Currently, there is a lot going on in terms of consultation and engagement, if not yet so much on collaborative planning, both in the UK and in other countries. There is very little in this book for which I can claim to be the sole inventor, in fact much in almost all chapters draws in particular on the further reading list in Appendix 5. Inasmuch as collaborative planning is about lots of different people coming together and sharing, so the same is true of the engagement community; the best ideas come from people working together, talking together and helping to comment on each other's work. As a result, many people have contributed to this book, some unknowingly (when it is not clear where an idea even started), some as a result of a courtesy call and some with substantial input, in which case the authors are credited or referenced. If there is any failure of appropriate crediting it is mine alone.

I would like to give my particular thanks for what has been a whole professional lifetime of stimulating and challenging thinking and action to (in alphabetical order) Andrew Acland, Lindsey Colbourne, Richard Harris, Rowena Harris, Allen Hickling, Pippa Hyam, Ray Lorenzo, Steve Smith, Penny Walker, Diane Warburton, Lynn Wetenhall and Cathy Williams (with apologies to those not listed). There are also many others from whom I have taken perhaps just one crucial statement, quote, idea or method: thanks to you, whoever you are. Thanks to Elinor Greenacre for the diagrams and figures. Special thanks are also due to Professor Patsy Healey whose wise guidance eventually persuaded me to write this book and helped to bring it to fruition.

1
Introduction

The consultation and engagement bandwagon

Nowadays, it almost seems that if anything at all is to happen in the public domain – small or large, specific or general, affecting many or perhaps being almost invisible – it must be 'consulted' upon. The pressure to do so has been building for some time and many governments have almost competed to be in the driving seat of the consultation bandwagon. Many professions have also now come on board the bandwagon, if to different degrees (although some claim to have been in favour of consultation all along). And, of course, there is 'the public' or 'the community', some of whom have been clamouring for a greater voice in decision-making for years, while others still have no interest at all until something is proposed near them (if then). Finally, there is a semi-profession of consultation managers or facilitators now beginning to emerge, keen to see their area of work become the norm rather than the exception.

This burgeoning world of consultation (see later for other terms) has also happened alongside endless government reports, legislation, general guidance, academic evaluation, critiques and case studies and a huge variety of practical guidebooks. So what is missing that this book adds, or what gaps does it fill?

Five gaps to fill

The terms used to describe forms of consultation and collaborative working have been changing over the last 40 or 50 years. Although 'participation' was the most common term in the 1960s and 1970s (see Nicholson and Schreiner, 1973), it is now rare in the UK although still used elsewhere. 'Consultation' is still used, especially in government material, but everyday conversation has drifted onto 'involvement' and the latest term in the UK is 'engagement'. This links to strands of work about 'consensus building' and 'conflict resolution' and other related strands about 'collaborative working', 'dialogue' and 'deliberation'. These all form a spectrum or hierarchy (see Chapter 2), yet the differences between them are not easily explained, even if they are important and argued over endlessly, and not just by academics. That suggests a first gap to fill.

Including the term 'collaborative planning' in this book's title was a key and conscious risk, mainly because it is poorly covered in the practical literature and far from being well developed in everyday practice. In addition, collaborative practice that is based mainly on intense, face-to-face work with relatively small, invited groups cannot legally be used in the UK as the sole way to advance or resolve a number of planning issues. The law in the UK requires at least some opportunity for involvement by all in many situations. At the same time, collaborative approaches have a strong academic pedigree, if more in the US than the UK (for example, Forester, 1999 and Healey, 1997). That is the second gap to fill (though from here on, other terms are used where appropriate, most commonly 'engagement').

This book is also needed because existing practical guidance is almost always focused on methods rather than overall processes. The diet, food and cooking analogy that runs through this book treats methods as 'ingredients', events as 'recipes' and overall processes as 'menus'. In just the same way that one cannot create a successful menu or recipe with a random selection of even the best ingredients, so one cannot deliver a successful collaborative or engagement process just by picking a few clever methods. The worrying lack of practical guidance on overall processes suggests an urgent need to fill this third gap.

The next point is about moving from one-off initiatives into approaches that are genuinely regular and normal rather than the exception, i.e. mainstream. Without making engagement mainstream one is always reinventing wheels, starting from square one again, building capacity that then gets lost and needs to be rebuilt afresh. Though most emphasis in this area has rightly been on building capacity in the community, real progress will only come when whole organisations make collaborative working or engagement the default setting and build in procedures, relationships, skills and budgets to deliver this. There is very little academic or practical literature that focuses on organisational change for better collaborative working, so this is gap number four to be filled.

Finally, it has been this author's privilege to work with a number of those who have been advancing practice on collaborative working, and all those other terms, in recent years and to work on projects showing how genuine progress can be made, as well as highlighting mistakes to avoid. As of now, there does not seem to be one source that brings all this experience together, hence gap number five to fill (or at least to begin to fill).

Why this title?

The words in the title of this book are 'craft', 'collaborative' and 'planning' and all are important.

The term 'craft' is used because this is mainly a book for practitioners and potential practitioners of collaborative planning (or consultation or... etc.). It will, however, also be valuable for those not leading or managing engagement processes but involved in some way. Even for professionals who never undertake such work themselves, it is important to at least know about collaborative working because eventual success depends as much on the knowledgeable support of all involved as on the skills of the deliverers.

The term 'craft' also suggests that collaborative working is not a science; ritually following the guidance is certainly no guarantee of success. But neither is it just an 'art', solely dependent on style, feelings and quality of relationships. 'Craft' is about blending care, rigour and clarity (science) with sensitivity to people (art). It might best be thought of as needing heart, head and hand:

- **Heart** because it is essential that those promoting and delivering it genuinely feel that all parties have a right to be involved and that the outcomes will be better if they do. If it is treated as just 'a job', people will pick this up almost subliminally and processes will fail.
- **Head** because there is a lot to understand, be aware of, be able to explain and deal with, plan for and deliver. In fact, it is the really hard and careful work behind the scenes that is most important in building success – one advocate says that 80 per cent of success is in the preparation.[1]
- **Hand** because it does not succeed by what one thinks or writes. It succeeds because of a lot of very practical work and, in particular, the use of the demanding face-to-face skills of working with people.

The term 'collaborative' is extremely important because the outcomes of the majority of consultation work, and even engagement, are still largely determined by those who initiate any process rather than those involved in it. This is often described, and almost always pejoratively, as a top-down approach determined by the few in power (see Chapter 2). That may seem self-evident because most initiators are public authorities or private sector companies and the usual recipients are the community or the public. Working collaboratively is, by contrast, about all key parties having full and equal opportunities to contribute and be properly listened to, about all sets of values, aspirations and cultural ways of thinking and working being valued and about outcomes that are widely, ideally fully, supported. This is what enables solutions to emerge that can be well beyond those expected or laid down by the initiator. Collaborative working is not about the banal replacement of 'top-down' with 'bottom-up' or about 'power to the people' but about processes and outcomes that result in what are often termed win/win solutions, even if that is usually rather like absolute zero: A standard to approach but perhaps never to actually reach.

Most examples of collaborative planning focus on processes in which a small but wide-ranging group of representatives (stakeholders) come together on a number of occasions to enter a deliberative dialogue to generate a widely agreed solution or plan. This raises two important cautions.

The first caution has been mentioned already and is rather blunt: The statutory planning context in the UK requires opportunities for involvement to be offered to all, not just a small, probably selective (even self-selected), group. The second caution builds on this because, although this book stresses the considerable added value from making carefully managed, intense collaborative working the *core* of any engagement process, the inclusion of some wider consultative work is also appropriate because it provides a further level of democratic legitimacy. That then helps to achieve broader social change, higher general levels of awareness and understanding, improved capacity for all and progress towards what is sometimes called participative democracy. Collaborative planning alone is challenging enough, but designing processes that also include wide-ranging public consultation is even more so. But that is an unavoidable challenge; another reason why the term 'engagement' is used, not as a substitute for collaboration but to embrace more widely inclusive approaches.

In everyday life the term 'planning' is used to describe organising anything from a family holiday to a national programme for healthy living. This book's main focus is unapologetically on planning in terms of the statutory, semi- or non-statutory processes of land use, town or spatial planning, so it is targeted at those who work or intend to work in that topic area. That therefore includes not just planners but also architects, engineers, project managers and others. As it happens, land use planning probably brings to the surface just about every issue or challenge generated by any other sort of 'planning' and seems to involve the largest numbers and widest range of people. So planning offers what is perhaps the toughest possible test of collaboration, certainly the most diverse.

Note also that the context and examples in this book are all from the United Kingdom, often just from England, because the detail of how engagement is commissioned, designed and delivered is inextricably linked to its social, cultural, economic, political and especially legal context. And that would apply in Australia, Argentina or Austria according to their own contexts.[2] Attempting to develop approaches that purport to be context-free would be a recipe for failure because any context is such a major factor in shaping good practice. Despite this, the basic principles, processes, methods and skills covered by this book almost certainly have some relevance to other countries and settings (for example, see Town and Country Planning Association, 2007).

This book's content also has much of value for policy areas other than planning, such as renewable energy, environmental conservation, waste management and transport; indeed some examples in this book are from these areas. These are also areas where the UK planning system's statutory requirement to consult very widely does not always apply, so solely collaborative approaches are possible.

Despite the practical focus, much of what is covered is rooted in theory or, more precisely theories (as there is as yet no single theory to cover everything in this book).

Content and coverage

Chapter 2 'Setting the Scene' starts by questioning why one would use engagement processes at all, arguing that a key first step is to make a conscious choice that engagement is appropriate. The chapter then outlines some of the forms, levels, principles and potential benefits of engagement and, briefly, the legal context of its practice in UK land use and spatial planning. The final part offers a medley of illustrative examples of successful engagement, mainly as an introduction to the detail that follows.

Chapter 3 is entitled 'Getting Ready'. Mistakes are commonly made by simply launching into a process before its managers are ready in terms of skills, experience, resources, basic information, time and so forth. Continuing the diet, food and cooking analogy, there is a need to check, very thoroughly, someone's dietary requirements before beginning to suggest a diet. In other words, there are questions to ask before setting out; about past histories, the scope (or not) for change in response to any results, who takes key decisions and possible threats or opportunities. One element of this questioning is particularly important: Identifying all possible consultees, participants or stakeholders. Finally, before setting off, it is important to be clear about the objectives of any engagement; essential for carrying any process through but also essential in order to look back and evaluate once finished.

This now takes us to the three key central chapters.

The first of these – Chapter 4 'Design to Deliver' – covers things as yet almost absent from the literature. Having gone through the getting ready stage, one can start to design an overall process, and 'design' is exactly the right word. The chapter outlines and explains all the main stages in preparing, agreeing and delivering a successful collaborative planning or engagement process. It also stresses the central importance of having such a clear overall process ('menu'), i.e. a well-balanced combination of events or activities ('recipes').

The second key chapter – Chapter 5 'Delivering into Detail' – moves on to some of the specific events ('recipes') and methods ('ingredients') that make up a good process. The word 'some' is used because there are so many basic options and a potentially infinite number of variations; in fact, almost any process should involve some variation of any standard model. The chapter moves from the design of events within a process to specific sessions within an event, each of these needing its own very specific 'recipe'. That is then followed by a section on methods ('ingredients'). The latter could take up a whole book so Appendix 5 refers readers to books that cover them better than is possible here.

Chapter 6 'Working with People' cuts across what is in the two preceding chapters. Continuing the cooking analogy, it is about key 'techniques' (making a roux or folding rather than beating), all as absolutely key to success as menu preparation, recipe making and ingredient selection. As already suggested, the cold ('heartless') application of guidance will not on its own guarantee success. A large part of success, especially for more intense collaborative working, rests with the way in which all involved engage with each other at a personal level. This is often considered to be about facilitation in the sense of managing people during an event, but it is far more than that. Establishing positive relationships needs to infuse everything, from the first chat with a decision-maker, through the wording of an invitation to the phrasing used in a final report. However, this chapter comes with the warning that no book can ever teach someone to be a facilitator; that can only come through practical training and real life experience.

Chapter 7 'Evaluating and Reporting' addresses two tasks for when an engagement process is complete. The first part is about evaluation, mainly once a process is *complete* but there is still much to be learned by undertaking some evaluation *during* a process. Evaluation can range from brief and informal to rigorous and independent, but some form of it is crucial. The second part addresses the preparation and circulation of some form of report or audit of the engagement activity. The more that legal status is given to engagement within the planning system (as seems to be happening), the more important such reports are becoming, and they and the processes they report therefore need to be proof against legal challenge. (And, of course, the alert reader will have picked up that any good report or audit is also in itself a form of evaluation, so these two parts could have been presented the other way round.)

Chapter 8 'Making it Mainstream' moves to a different level. It is one that few readers are ever likely to be able to act upon on their own but it is extremely important. It can again be thought of in analogy terms as overall 'cuisine', i.e. a whole culture and way of cooking that is distinctive to a country, region or city (French country or Milanese style). Operating good, one-off collaborative or engagement processes is valuable but in any given area several engagement processes will always be underway at one time. Do they use common principles and agreed stakeholder lists? Is anyone checking that processes do not overlap or repeat? The sad answer is usually 'no'. To achieve real progress, all engagement practice in any area should be coherent, programmed, consistent and managed, helping to build capacity, skills, confidence and experience so that the whole system is a step up every time a new process starts. That requires an organisation-wide understanding and commitment, a whole culture of engagement. Unfortunately, this is as yet very poorly developed so the chapter can only begin to move things on.

Chapter 9 'Conclusions or an Engagement Utopia?' is hopefully more useful than the usual repetition of key points because it offers a slightly tongue-in-cheek story about an imaginary place where all of the key messages from all the main chapters have miraculously been taken on board.

There are then several Appendices about signposting and offering the reader further information on, and links to, some of the background theory about collaborative planning, the legal context for consultation, resourcing engagement, suggestions for how to break into the engagement territory and suggested further reading.

The chapters are not strictly sequential. There are almost always occasions when it is not just appropriate but actually good practice to treat things iteratively, to go back to thinking again or more deeply about the scope of a process, who to involve or methods to use. There is a constant need to refer back to even the most basic aspects as a process evolves, to not just go from one stage to the next without checking back and perhaps involving a change to the approach.

Throughout Chapters 2 to 7, a single, continuing example is used to illustrate the points being made. The example is from one of the author's projects and it is certainly not a shining example of everything in the book being delivered successfully. In fact, it was chosen precisely because there is as much to learn from its failures as from its successes. For this reason the names in this example have been changed. The invented name of this place is Waterton and its story is in the shaded boxes. Where other examples are used, actual place or organisation names are used, except where there is a need for anonymity to protect those involved, especially in those cases where often quite tense work is still continuing.

Finally, a small warning. This is not a quick 'tips and hints' book; it does not offer instant, easy answers and lots of clever diagrams. That is not an apology because it would be almost insulting to suggest that any of the challenges around environmental change and development could be tackled by a clever method here or fun technique there. This book digs deeper so that the reader will understand better *why* certain things happen, *why* stakeholder A might be relevant but not stakeholder B, and *why* method X would be appropriate rather than method Y. Perhaps the classic analogy for anyone running a successful engagement process is that of the swan: All calm and quiet above water, while all the hard work is going on below the surface!

Notes

1 Thanks to Allen Hickling for this powerful statement.
2 The term 'collaborative' does, however, have worrying connotations in Italy, for example. As this author was told: "we shoot collaborators"!

References

Forester, J. (1999) *The Deliberative Practitioner*. Cambridge (Mass.): MIT Press.
Healey, P. (1997) *Collaborative Planning: Shaping places in fragmented societies*. London: Macmillan.
Nicholson, S. and Schreiner, B. (1973) *Community Participation in City Decision Making*. Milton Keynes: Open University Press.
Town and Country Planning Association (2007) *Community Engagement in Planning – Exploring the way forward*. Final Report of the INTERREG IIIB Advocacy, Participation and NGOs in Planning Project (APaNGO). Available at: www.tcpa.org.uk/pages/apango-reports.html

2
Setting the Scene

Introduction

Some years ago, a planner was asked to consult with 'the community' to find out what they wanted to see developed on a derelict site. He hired a tent to put on the site a couple of weeks later and promoted a community event. He and colleagues ran the event from a Friday evening to a Sunday afternoon. Well over 400 local people called in, made notes on maps, listed issues and suggested ideas. Over 100 people were present for the final session, at which there was clear agreement to develop the site for housing for local people and a community building. People were excited and engaged, so the planner was asked what would happen next. Things now fell apart. First, he had not thought beyond just one fun event; he thought that was all consultation was about. Second, he had not noticed that a significant number of people contributing to the event were not local, coming instead from a city-wide social housing company. Third, it was only when he shared the proposals back in the office that he discovered an already well advanced scheme to use the site for offices. Unsurprisingly, the reaction from those living near the site when they heard about the office project was one of extreme anger!

Consultation (or engagement or collaborative working) is not something that is just 'fun' (though many methods can or should be). Nor is it something to launch into without considerable thought about an overall process; it is not something that can be done in one event or separate from the actions of key organisations. In fact, the first step when considering any type of engagement activity is to take a well-argued, well-evidenced, carefully thought-through decision as to whether it is even appropriate. There is, in other words, a sort of pre-principle of engagement: *Know when not to!* As illustrated in the aforementioned story, engagement is too often started at a time that is inappropriate, which brings practice into disrepute and lessens community confidence and trust.

Is there really such a blunt division: Do it properly or don't do it at all? There is some truth in this and if this book was solely about pure forms of collaborative planning then that is where things might stop. However, not only is collaborative planning in the real world not pure and perfect but it is also part only of a spectrum of approaches, even if it is significantly richer and deeper than most.

If some form of collaborative working or engagement *is* appropriate, this chapter sets the scene by outlining:

- The different **forms and levels** of engagement.
- Its potential **outcomes and benefits**.
- Some **overall principles of good practice**.
- A few key **principles of process**.
- Some basic **legal and quasi-legal requirements**.
- The main **barriers** to making engagement the rule not just the exception.

There is also a final section offering examples to show that it has real value if, of course, it is done properly.

Forms and levels

Various terms have already been used in this book. This is deliberate because 'collaborative planning' is meaningless in most community settings, and not always useful in professional settings (some may mouth it, few can elaborate it); it is a term that may never gain common usage.

The words 'consultation', 'involvement' and 'engagement' have been used so far and are in regular use today in the UK (see, for example, Great Britain, 1994). The point was also made in Chapter 1 that the term 'participation' was used for many years but is rarely heard nowadays, in the UK at least. Both 'dialogue' and 'deliberation' describe how collaborative working takes place and perhaps the final term to add to this core list would be 'consensus building' (often linked to its darker side of 'conflict resolution').

Defining and differentiating between these terms has kept academics in books and conferences for years, without as yet any clear agreement except at the broadest level. Practitioners and even some academics now find themselves getting a little closer once a spectrum, hierarchy or a range of levels is introduced. All of this makes three things clear:

- The dividing lines between levels are uncertain.
- It can be dangerous, even wrong, to consider so-called *higher* levels to be *better*.
- Good processes can, even should, include opportunities for different people to engage at different levels at different times and on different aspects.

Here are three alternative models, or are they just variations?

1. Creighton asked (Creighton, 1992), when participation was still the usual term, "what does it take for a decision to count?" and introduced four ways in which the public could have an influence:
   - **Public information**: To be informed of the decision.
   - **Formalised participation**: To be heard before the decision.
   - **Consultation**: To influence the decision.
   - **Consensus building**: To agree to the decision.
2. A UK organisation called Involve outlined five levels of what they term 'participation' and unusually used 'participation' in the title of their book (Involve, 2005):
   - **Inform**: To provide the public with balanced and objective information to assist them in understanding the problem, alternatives and/or solutions.
   - **Consult**: To obtain public feedback on analysis, alternatives and/or the decision.
   - **Involve**: To work directly with the public throughout the process to ensure that public concerns and aspirations are constantly understood and considered.
   - **Collaborate**: To partner with the public in each aspect of the decision-making, including the development of alternatives and the identification of the preferred solution.
   - **Empower**: To place final decision-making in the hands of the public.
3. Very deliberately going back to just three terms in a more recent but not formally published report, Colbourne and others suggest the following:
   - **Transmit**: To inspire, inform, change, educate, build capacity and involvement or influence the decisions of stakeholders.
   - **Receive**: To use the views, skills, expertise and knowledge of stakeholders to inform, change, educate, build capacity and involvement or influence decisions.
   - **Collaborate**: To collaborate, consider, create or decide something together with stakeholders.

There are many other versions of the first and second models described here but all, almost without stating it, imply that higher levels are better. The third model, however, steadfastly refuses to suggest that one is better than the other; it implies a 'horses for courses' approach and that any process probably needs some activity at all three levels.

More importantly, when considering collaborative working, almost all imply some agency whose members or officers decide when any engagement will take place, when and how to initiate it and the level at which, style in which and activities through which it will be delivered. Put another way, they decide what scope the public or the stakeholders (a term explained in Chapter 3) have in any situation. The use and scope of participative or engagement approaches is therefore most usually at the discretion of those in power, and power and its distribution are unavoidably central to any discussion of engagement practice. There is an old, endlessly recycled joke about this, the origins of which are now lost, and it is hardly a joke as it remains too often true. It asks how one might conjugate the verb participate, to which the answer is:

> I participate, he participates, she participates, we participate, you participate, … *they* decide!

At this point it is essential to introduce an almost classic set of levels, certainly within the world of land use planning. This was devised by Sherry Arnstein in the USA in the 1960s (Arnstein, 1969). It can now be found in academic literature and practical guides in dozens of countries and languages. Arnstein's concept was of a 'ladder of participation' and, unlike some of those mentioned, she was quite explicit that the upper levels were better; in fact the very terms she uses makes that obvious. Here is the original ladder, see Figure 1.

Arnstein's concern, typical of the 1960s, was for wresting power from the powerful and transferring it to citizens. In many ways her ladder remains a strong and useful analogy. For our purposes here, however, it has one serious weakness that must be addressed if any form of collaborative working is to move forward, highlighted by looking at the ladder in another way, see Figure 2.

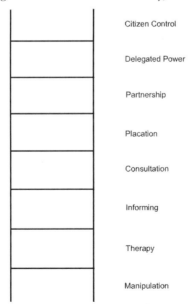

Figure 1 Arnstein's original ladder

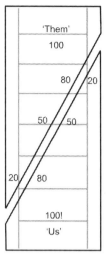

Figure 2 Another way of seeing Arnstein's ladder

This shows that a shift up the ladder merely places more and eventually all power with one party or another, from 80/20 lower on the ladder (more to those in power) to 20/80 near the top (more to the citizens) and 50/50 in the middle. This is fundamentally a win/lose model and, for any complex social issue, it is probably impossible to argue that the wholesale transfer of power from one group to another is any form of progress at all. By contrast, collaborative working and consensus building are about trying to achieve a win/win result in which there is real added value from bringing together skills, knowledge, information, power and so forth from a range of parties. It is about both or all, not one or the other.

In Chapter 1 it was suggested that collaborative working that achieves full consensus is rather like absolute zero; good to aim at, probably never reached. That is undoubtedly the case, and there are some academic books that undermine the claims of those who pretend that genuine, full consensus has ever been achieved. Their critiques are most commonly about the failure to properly shift or erode traditional power bases in, for example, government and the professions, about the way in which the language and methods used are not really open and equitable for all, about the failure of some (the powerful again) to accept and act upon emerging agreements and about the nature of one-off successes which nevertheless fail to change overall policies or organisational cultures (see Chapter 8).

This book is based on the principle that there are crucial and valuable advantages to be secured from setting out to engage all at some level, from trying to avoid predetermination of the outcome by any initiator of a process and from taking time and care to get as close as possible to agreed, consensus, win/win solutions.

Having suggested that there may be no perfect typology, this author has contributed, along with colleagues, to one that seems to go down well with a very wide range of people (professional and public) and avoids some of the traps of others (e.g. higher levels being better). Table 1 offers a four part basic framework (the left hand column) but takes this further. It explains a little about each approach in the second column and then gives examples of the type of method one might use for each approach.[1] Finally, the arrows to either side show how certain approaches *increase* influence (arrow to the left) while also at the same time probably *decreasing* the number of people likely to be involved (arrow to the right). This is a key challenge for those promoting only small group collaborative approaches.

Outcomes and benefits

Having outlined what collaborative working is about, what outcomes and benefits might it deliver? Thorough engagement and collaborative working undoubtedly demand a lot of effort from all, so do they deliver anything that could not be delivered by traditional consultation, or by doing nothing (which at least raises no expectations)? Proponents suggest that good engagement can:

- Secure wide agreement on the purpose and direction of a project, programme, plan or policy.
- Introduce more detailed knowledge, skills, issues, experience, ideas and judgements alongside, not instead of, those of professionals.
- Draw out and clarify, perhaps prevent but also diffuse or resolve conflict.
- Speed the overall timetable.
- Use all resources more effectively and probably save overall, including in terms of money.
- Improve the direct or prime outcome: A better, more sustainable, deliverable, affordable plan, project or policy.
- Add value through indirect benefits such as feelings of ownership, increased acceptability and speedier implementation.
- Increase involvement and encourage mutual education, understanding and respect.
- Improve personal and working relationships notably between the more powerful and the less powerful, professionals and lay people.
- Build confidence, skills and trust for next time, growing social capital and building capacity for all.
- Establish (perhaps re-establish) trust in governmental processes and systems and in principled professionalism.

Table 1 A preferred levels framework

Type	Description	Example methods
Informing	Letting others know of decisions, opportunities, ideas. May be about addressing attitudes/perceptions or behaviour (through education). Informing may also involve sharing – listening to different points of view, and allowing people to understand differences, rather than explicitly trying to inform about decisions.	• Giving information through leaflets, guided tours, websites, e-news • Running education programmes • Giving talks • Public Relations work through the press
Consulting	You will make the decision, but you want to inform that decision by gathering stakeholders' views first. Often one-off engagement.	• Market research surveys • Focus groups • Exhibition/questionnaires • Public meetings • 1:1 meetings • Face to face or electronic • Citizens' Panels (post/email)
Involving	You will make the decision(s), but you want stakeholders to be able to shape decisions on an ongoing basis. This results in longer term and more influential relationships in which final decisions are made by the commissioning organisation(s), but very much based on the working relationship with those involved.	• Advisory bodies • Liaison groups • 1:1 relationships with owner occupiers • Workshops • E-consultation • Citizens' Panels (meeting)
Dialogue	You share and develop the decision-making more equally with your stakeholders on an ongoing basis that allows for and balances all views.	Partnerships Consensus building Managed delegation E-dialogue Action packs

Increasing influence/control over decision →

← *Increasing number of people involved*

No single project or process is ever likely to score 10/10 against the above but working collaboratively is probably the only way to achieve some of the things on the list. Narrowly defined, instrumental consultation targeted solely at (for example) securing planning permission cannot provide genuine agreement, enable and balance the widest range of inputs or maximise resources. Most importantly, it cannot deliver the benefits of changed feelings of ownership, improved working relationships and enhancement of social capital, all key benefits in the longer term and beyond the single project.

Overall principles

Although there are probably as many lists of principles of engagement as there are books on the subject,[2] there are few as yet that are described as principles of *collaborative* working. As with outcomes and benefits, it is possible to combine lists together to come up with a generic version, though none can ever be right. What follows is in a broad sequence from starting a process through to its finish. This is the author's personally preferred list (so it is no more right than any other), but it has proven itself through using it with a very wide range of people.

- **Explicit overall process**: Success comes from the careful, thoughtful and explicit design of a coherent, cumulative, overall process (or 'menu' using the cooking analogy). That process needs to be transparent and clear to all.
- **Independent design and management**: Any process should be designed and managed with demonstrable independence from those commissioning and funding it. Independence is never an absolute but it can still be demonstrated to be, and accepted as, *appropriately* independent.
- **Agreed process**: The design of a process should ideally be agreed with at least some of those other than the main commissioners, typically with some key stakeholders. Delivering an agreed process can significantly enhance the chances of demonstrating independence, securing wide final support and decreasing the scope for later challenge.
- **Clear scope**: In most situations there will be fixed issues, some that might be changeable and some that are certainly changeable. It is an essential prerequisite for trust to make these clear at the outset because it is dangerous, almost insulting, to suggest that there is ever a blank sheet and everything is open to change. Anecdotal evidence suggests that people are happier to be given a clear if limited scope to their involvement and then see some small changes than they are to be offered apparently unlimited scope and then see nothing change. Defining the scope clearly at the outset also enables those involved to challenge that definition. (This is a classic

example of knowing when not to; if there really is nothing that can be changed, *do not consult!*)
- **Commitment to abide by outcomes**: This is important but also challenging because the outcomes of collaborative processes are never definable at the outset. Ideally, all involved, especially those with ultimate power of approval, should commit at the outset to abide by whatever emerges from the process if that process is delivered properly. If not, that needs to be clarified under 'scope' as above.
- **Openness, honesty, trust**: Commitment can, however, only reasonably be secured if all parties are open and honest, and if the process builds trust through communication based on two-way listening and questioning and an exploration of needs rather than positions. But even if things are genuinely open and honest from the outset, trust can take some time to build, or rebuild if previous experience has been damaging.
- **Inclusiveness**: Collaborative working demands great care in establishing, as early as possible, the whole range of possible views about and interests in an issue/project, and involving all relevant individuals and groups; not just the obvious friends or even enemies, not just the locals, not just the professionals. Any process must also offer genuine opportunities to contribute and on people's own terms.
- **Mutual education and exchange**: If information, attitudes and values are in the open, and shared between all at all stages, there will inevitably be a shifting of perceptions and a development of personal and group knowledge. There is inevitably an educational dimension and that is not just for the public, the community or lay people; it is as much for professionals, elected representatives, commissioners and others.
- **Multiple options are identified**: It is banal to suggest that any complex problem has one neat, simple solution. A range of people approaching things from different directions can generate diverse and innovative options and solutions and a base for a more creative, agreed solution which is a genuine choice. Keeping options and choices open for as long as possible is important because it helps to build trust and confidence that the finally preferred option really is the best.
- **Building common ground**: Although the most difficult problems require the most attention, it is important to seek out and build on those points of agreement and common ground that nearly always exist or can be developed quite quickly. Some may be minor, but acknowledging small steps helps to create confidence and mutual trust, providing a platform from which to move to more challenging issues.
- **Common information base**: Conflicts often roll on simply because different groups argue from different bases of ideas and information. A consensus process pays attention to sharing and valuing all information, seeking

agreements about the information, and seeking further agreed information to take things forward. Clearing the 'fog' caused by misunderstanding or disagreement about basic facts helps to focus on those issues genuinely requiring attention.

- **Methods not method**: Any event or method ('recipe' or 'ingredient') can only be as good as the overall process ('menu') in which it is used. Within limits, methods are not good or bad in their own right. In addition, no process even on an apparently simple project can ever be successful if reliant on one method alone. Different people respond in different ways at different stages to different methods or techniques. If chosen and managed with care, methods can also work corroboratively, mutually reinforcing each other to give more robust results than can be achieved by, for example, a larger sample number from a single questionnaire.
- **Decisions made by consensus**: Many procedures still rely too heavily on the ultimately debilitating system of majority votes. Either that or someone at the centre of a process makes up his/her mind solely on the basis of whatever evidence has come in (this is often termed 'black box' decision-making). Working towards decisions which are shared with many and eventually supported by the largest possible number greatly increases the chances that others will back, rather than prevent or delay, approval and later implementation.
- **Clear feedback and final reporting**: Far too often, people enter participatory processes and never hear what happened to their contribution. It follows directly from all the above that everybody involved should receive not just results but enough information to provide a robust audit trail back to anything they and their colleagues have contributed. This applies even if, perhaps more so if, the outcomes are awkward or challenging or the process has failed to reach overall agreement.
- **Shared responsibility for success, outcomes and implementation**: Once collaborative work is underway, those involved are not passive actors waiting for the mythical 'someone else' to solve things for them. Everybody should accept responsibility and an active role in seeking progress, actively supporting the process and backing the eventual outcome (and how it was reached). It is also important that any support goes right through to final implementation.
- **Work hard on the detail** (this applies throughout; it is not sequential): Abiding by all of the above can be completely ruined by poor choices of venue, poor invitations and briefing, badly chosen dates, not having enough materials, not providing refreshments, failed technology, poor time-keeping, lack of clarity, not reporting back and so forth. It is the hard, detailed and often invisible backstage work of preparation, reporting and more that generates real success (see Chapters 5 and 6).

This list may be important but it is also long, so here is a summary of the main titles:

Table 2 Principles of engagement summarised

- Explicit overall process
- Independent design and management
- Agreed process
- Clear scope
- Commitment to abide by outcomes
- Openness, honesty, trust
- Inclusiveness
- Mutual education and exchange
- Building common ground
- Common information base
- Methods not method
- Decisions made by consensus
- Clear feedback and final reporting
- Shared responsibility for success, outcomes and implementation
- Work hard on the detail

Principles of process design

The principles described are mainly about *what* needs to be done rather than *how* it might be done, and the word 'process' is used a lot. Although process design is explained in detail in Chapter 4, three design principles are so important, and affect so much that follows, that they need to be introduced here.

Needs not positions

Some (most?) societies can be described as inherently adversarial in that the default mode for most people in most situations is to start by taking a position and defending it against assumed attack by others (see Fisher and Ury, 1981). This is, of course, exactly what the other party will be doing too. We see this at play every day in the courts, public inquiries, council chambers and Parliament.

This is classic win/lose thinking, based on the assumption that there is only ever a certain amount at play in any situation and either I get all or most of it or I lose and you get all or most. It is also based on the assumption that if I come clean too early about what I am seeking (my bottom line) you will take full advantage of me and I will lose out more radically. All of this is of course self-reinforcing such that, once set in this mode, it then 'proves' that opening up will only ever lead to some sort of loss, and so on and so on, forever spiralling down!

Consensus processes cannot resolve things if those involved are stuck in adversarial mode; the way forward is to get them out of that mode. That requires finding ways to enable them to dig deeper into themselves, their situation, the situation as a whole, what others bring and what is potentially on offer, and feel able to share this more openly. Rather than taking positions, the key is to start to draw out people's interests and, if possible, what they really need from a situation. Only once interests and needs are clarified and in play can a creative resolution – a win/win – be achieved. This can be summed up in Figure 3.

- The top of the diagram shows that taking *positions* inevitably involves a narrowing down of needs to produce something defensible against attack and this leads, almost inexorably, to conflict; for example, taking the fixed position that new development can only ever lead to increased traffic and associated problems.
- *Interests* can be more personalised but can also be broader and may well be shared between several parties in any situation; for example, an interest in traffic movement at a particular junction.
- *Needs*, however, are rarely the domain of a single person and are more likely to be socially shared; for example, improved road safety for children around a road junction.

Once into the area of needs and some sharing of aspirations, there is scope for genuinely mutual benefit.

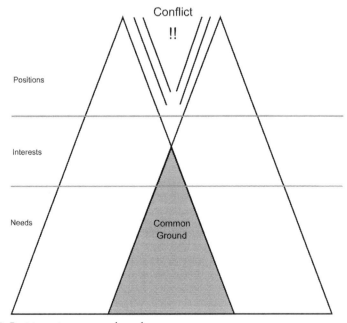

Figure 3 Positions, interests and needs

Setting the Scene

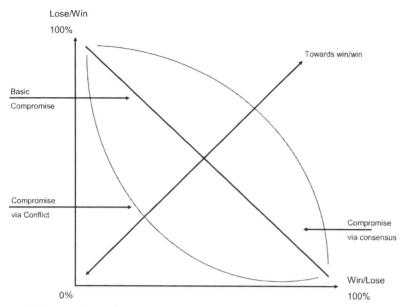

Figure 4 Compromise and consensus

There is another way of looking at this because true win/wins are very rare. If two people are involved in some form of negotiation there is little chance that they can both get 100% of what they want or need. But neither should they be stuck with just managing a 70/30, 80/20 or even 50/50. In the diagram shown in Figure 4, the straight diagonal illustrates the simple win/lose line.

As the diagram implies with the 'Compromise via conflict' line, the angst, tension, time and cost of traditional conflict processes almost always lead to a less than optimal outcome; not even 50/50 but more like 30/30. This is easily thought of in terms of the very many potentially valuable plans or projects that might well have happened but did not, *solely* because their processes of development were so conflict-based. Consensus approaches might not be able to manage 100/100 but they can add value and create something like 70/70, as with the 'Compromise via consensus' line.

Getting the sequence right

One damaging outcome of adversarial thinking can be seen right at the heart of traditional consultation processes. This has been characterised by one author and consensus building practitioner[3] as processes of 'Decide–Announce–Defend' or just 'DAD', as in the diagram in Figure 5, the horizontal line showing time.

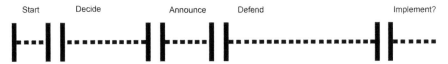

Figure 5 Decide–Announce–Defend

In this approach, those in power and in control of the process start by sitting in a closed room not just to design the process by themselves but also to draw up a proposal by themselves. During this *Decide* stage they do not talk to or engage others, those who might be impacted by the proposal, and all those with a potential contribution to make. Those others remain either unaware of the process or know it is going on but cannot affect it. Yet, those in charge probably know not only who is likely to object to their proposal but could even write their letters of objection for them!

And then the closed room proposal is unveiled. The vast majority of people will neither know nor (at that stage) care, while those who did know or are now concerned respond to the proposals by protesting about exactly what those in control knew they would protest about. This is the *Announce* stage, which can take some considerable time (and money) because nobody knew the proposals were coming. At this point the core team move into the *Defend* stage, made easier by the fact that their proposals are no doubt beautifully presented in expensive documents, full of technical jargon and not readily available to the public, plus the fact that they, and only they, are party to the information used to produce the proposal. And, almost inevitably, the other people now have only a very limited time in which to respond!

This is, however, just the start of the *Defend* stage because it can and does go on and on, from informal responses to statutory inquiries, to appeals, to the courts and so on, and all at huge expense in every sense and with enormous frustration. And finally, because it has been so badly managed, the plans may never go through; not even a 30/30 result but a 0/0 one if the proposals fail.

The alternative, consensus-based approach is called 'Engage–Deliberate–Decide', or 'EDD', but is more commonly referred to, based on one of its distinctive features, as 'front-loading', i.e. bringing people together from the very start. The equivalent diagram for EDD is shown in Figure 6.

Figure 6 Engage–Deliberate–Decide

As the diagram shows, getting started takes longer because there is value (as in the principles above) in engaging others to agree the process. Then the *Deliberate* stage takes longer than the *Decide* stage in the DAD model because it is about involving as many as possible in a more collaborative approach to generate a widely understood and widely agreed proposal. Some *Announce* activity will then still be necessary because not everybody will ever know that a process is under way. However, not only will far less of an announcement be needed, because many will already be aware as a result of their involvement, but many of those will themselves not just help with the announcing but will make it clear that they support the plan, indeed some are likely to 'champion' it.

The experience (see the final section of this chapter) is that this approach reaches better and more agreed proposals and that it does so more quickly and more cost-effectively in the end, even if the early stages take longer and perhaps cost more. By using the EDD approach, more proposals are then likely to end up being implemented rather than scrapped; itself a serious financial benefit.

Inclusiveness

This is one of the key principles of engagement previously discussed and mentions have already been made of engaging 'everybody' or 'all'. The practicalities of listing out who 'everybody' might or should be are covered in Chapter 3 but for now there are some key points to make about inclusiveness.

Though it is very, very rare to ever be able to involve everybody, there are two well-known patterns, as shown in Figures 7 and 8, to the likely levels of practically achievable involvement. The first is *size of area*, which can be expressed in a simple graph, see Figure 7.

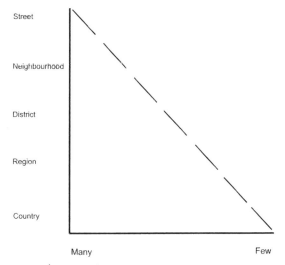

Figure 7 How many might engage?

From this author's experience in his own community, it *can* be possible to get every resident in a street involved in a project, in this case tree-planting. By contrast, anybody trying to engage people across a whole county in a waste management strategy would be pleased to get a 0.01 per cent response rate to a questionnaire and perhaps 50–60 stakeholder group representatives at a workshop. The size of the area and the number of people potentially involved have a significant and unsurprising effect on involvement rates.

The issue of size also relates almost one-to-one with an issue about *immediacy*, i.e. the extent to which a plan or project, or some element of either, is seen by people to have an immediate or potential impact on them. That generates a graph similar to the one shown in Figure 8.

The tree-planting project had clear immediacy for everyone living on that street. By contrast, many in that same street found it difficult to understand the relevance of the Council's waste strategy. In fact, only two people commented on the part of the waste strategy that proposed to introduce a variety of refuse bins, but many were up in arms once the bulky and rather ugly bins arrived. A waste strategy might mean little to most but a lot to some if it suggests a location for a composting site, for example. However, proposing an incinerator could provoke reaction from a very large number, including some at national, even international level. Certain developments might seem innocuous to many but be seen by some as likely to devalue their houses or affect their whole lives. All of these things affect the number of people likely to wish to have a 'voice' in an engagement process.

With just these two factors in mind, debating who to involve in a process is a difficult decision. In other words it is never possible, maybe not even appropriate, to consider accessing all. What is more, different people or groups

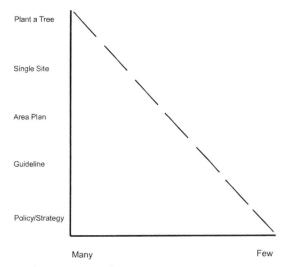

Figure 8 How immediate is any issue?

will wish and/or be able to engage to different degrees or on only certain aspects; for some that will be minimally at the start, for others in depth throughout and for others just to check towards the end about their specific concern.

Collaborative processes are most commonly undertaken with limited numbers of representatives (stakeholders) because there is a need to spend enough time to get people past positions and into interests and needs, to explore and develop often considerable amounts of information, to take key points from key stages back to a wider group, to develop and evaluate options and to agree end results. Even if methods were available to enable huge numbers to engage in face-to-face dialogue, many would still not wish to do so, either just wanting or being able to dip in only briefly, if at all. Once again, there is no value in arguing that one format, level or stage of involvement is right. What is right is enabling all to understand enough about any issue early enough to be able to decide what involvement they wish, and then for that to be offered and delivered.

One way to resolve this is to avoid reliance on one particular approach or method (as in the principles). This can have real value as illustrated in Figure 9, which makes a rather simplistic contrast between methods to engage a small number very intensively, here termed 'in-depth', and methods targeted at a lower level of involvement by all, here termed 'in-breadth':

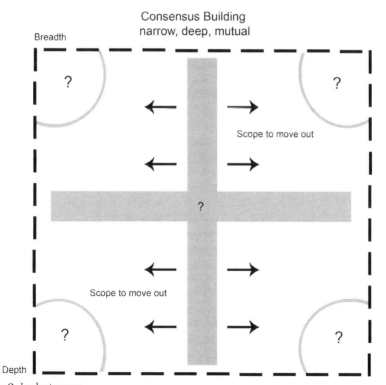

Figure 9 Inclusiveness

- The line around the edge suggests all those who may need to be involved but it is a broken line because some may come in once a process is underway and some may leave.
- The vertical band refers to the relatively small number of stakeholders who may be involved throughout, almost certainly by careful invitation and therefore in-depth.
- The horizontal band refers to all those who may only wish or be able to get involved minimally (in-breadth), but what is also implied is that, although this opportunity may be offered to all, only a small proportion may take it up (for example, with a questionnaire).
- The areas in the corners refer to those often called the 'hard to reach', an issue dealt with in Chapter 3.

As explained in Chapter 4, careful combinations of work in-depth and in-breadth add value to any process and deliver benefits beyond those possible with either breadth or depth work alone. This is particularly true because those involved in-depth often have a greater ability to reach, persuade and engage others who would otherwise never get engaged, as suggested by the arrows out from the vertical band in the diagram.

The legal and quasi-legal context

This section requires two cautions. First, a repeat of the caution raised in Chapter 1 because the text (and almost all the examples in this book) is solely about the English planning system.[4] Some other countries operate rigorous legal systems for engagement in planning, some have none at all. The key point is that consultative and collaborative practice must relate to its own legal context. Second, all of what follows, indeed this entire book, is about the early stages of what is generally described as *informal* consultation. However, both in plan-making work and for planning applications there are also stages of what is termed *formal* consultation. These are narrowly defined and offer very little opportunity for change (see Appendix 2).

A brief history

It was in 1947 that the UK's first legislation requiring some sort of participation in land use planning appeared. Looking back, that requirement barely scratched the surface of participation as outlined in this book. It was solely about a requirement on planning officers to consult neighbours on planning applications, report any responses and use the responses in shaping their reports on those applications.

Forms of 'participation' (the term at the time) were developed further during the 1960s and 1970s, reflecting that period's more radical social context. Arnstein's work has already been mentioned (Arnstein, 1969) and, around the same time, a more formal yet also challenging set of ideas emerged in a report entitled 'People and Planning' (Skeffington, 1969). This generated some interesting, quite forward-looking and often successful experiments and initiatives, several supported by the government of the time, but they were short term and entirely ad hoc. However, little of any of this managed to find its way into legislation; the real shift started in the early/mid 1990s from two seemingly contradictory directions.

First, the government of the time commissioned research on 'Community Involvement in Planning and Development' (Great Britain, 1994), led by this author. This produced evidence showing clear benefits from undertaking coherent involvement in both plans and projects and offered a framework of good practice. However, having printed thousands of copies of the report, the same government refused to allow it to be publicised!

Second, at exactly the same time, a now defunct national agency (the Countryside Commission) commissioned a study aimed at improving design standards in rural areas, again led by this author and a colleague. One of the approaches that emerged was called Village Design Statements. These were to be prepared and even published mainly by communities themselves, if with planner input, so they were a heavily delegated, 'bottom-up' approach. This was probably the first time in the world that local people could produce documents that then became a part of the suite of formal, statutory plans (and over 1,000 have been produced). Strangely, the same government that buried the report as above fully supported, even celebrated, Village Design Statements!

Between then and now (autumn 2014), much of the above has resurfaced and this time in legislation, first under a Labour government and then a Coalition one (of Conservatives and Liberal Democrats).

The 2004 changes

The first major set of changes, mostly retained by the Coalition, was Labour's Planning and Compulsory Purchase Act of 2004 (Great Britain, 2004). Better engagement was a major theme behind this Act (Great Britain, 2004/2) and one key new element was 'Statements of Community Involvement' (SCIs).

A Statement of Community Involvement was legally required to be produced by each local planning authority, laying down in broad terms the overall standards to be used, aspects to be covered, who must be consulted and procedures to be followed for 'community involvement' (note the words) during plan preparation. Once adopted (the formal term for legal support), the SCI was to be used by the authority planning officers to guide its processes.

This was all backed up at the time by a medley of guidance from, funded by or endorsed by government (for example, Planning Aid, undated).

The term 'front-loaded' first appeared here, generally endorsing the Engage–Deliberate–Decide approach; for example, Regulation 25 requires that local people be involved in *developing* the key issues, in *developing* the information base and in *developing* the basic plan options. This does not mean minimal consultation; it implies active and early engagement alongside the professionals. Once plans were prepared they were, as before, to be submitted for formal examination accompanied by a report (rather confusingly also called a Statement of Community Involvement) showing how the standards of the SCI had been fulfilled.

This was potentially a huge leap forward but the government's own monitoring research (Great Britain, 2007 and 2008) suggested that the reality had fallen far short of the ambition. Far from being front-loaded, early plans were most commonly still prepared through Decide–Announce–Defend (DAD) approaches. In fact most planning teams still produced initial 'Issues and Options' reports (hence using a supposedly out-of-date term) prepared solely by themselves in darkened rooms. This cannot have been helped by the fact that, during formal examination, there did not appear to be any rigorous analysis by Planning Inspectors to check, against an authority's original SCI, what had been done and how the results had been used. This message got back very quickly to local authorities where highly pressured staff chose, perhaps understandably, not to commit resources to front-loaded involvement if it was not to be valued at examination.

SCIs were also required to include encouragement to potential applicants for planning permission to undertake what is termed 'pre-application involvement', however, the government baulked at legally requiring this. Though often not stated very clearly, the same general standards about plan preparation were supposed to be applied to pre-application work. In general, this made no difference at all to actual practice. Some applicants, notably government agencies and a few developers, already undertook pre-application involvement and carried on doing so, but very few started afresh as a result of what was stated in any SCI.

The 2010/2011 changes

If the 2004 Act picked up from, amongst other things, the 1994 government research mentioned above, the Localism Act (Great Britain, 2011) drew significantly from Village Design Statements and several other bottom-up, community level initiatives.

In relation to the points above, it is important to note that the Localism Act took one step forward by stating that pre-application community involvement would become a mandatory requirement (via secondary legislation), if probably only for larger projects. As yet there has been no progress with the secondary legislation and, given earlier experience, there must be concern about what weight might eventually be given to the outcomes of pre-application involvement. It is, for example, very doubtful that councillors (or planning officers) will ever feel confident enough to use poor or even no engagement as the sole reason to refuse a planning application.

Having announced in the Localism Act their intentions for pre-application involvement, the Coalition shifted tack in the National Planning Policy Framework of 2012 (Great Britain, 2012). In at least two places in the text, support was given to pre-application consultation on *all* projects, especially on design issues, and there was even a strong statement (paragraph 66) to the effect that "proposals that can demonstrate this in developing the design of the new development *should be looked on more favourably*" (author's emphasis). The Framework has the status of law but again, without the secondary legislation, even this strong statement is unlikely to carry any weight.

The other innovation in the Localism Act was not directly about engagement. Neighbourhood Development Plans (NDPs) are better seen as a link between what had been started by Village Design Statements and what was in higher level plans, being a mainly community-led format for land use plan-making. They are important because they introduce another level in the hierarchy of statutory plans and one that is not inherently top-down as are all others. Neighbourhood Development Plans are required to go through an examination process and must be accompanied by a report of community involvement. Once a NDP has been examined it is then adopted (technically 'made') if it secures over 50 per cent support in a neighbourhood referendum.[5]

It is too early to pass judgement on NDPs, including their level and quality of community involvement. The first NDP to plan for major development (for Thame in Oxfordshire) was produced through an engagement process led by this author which, according to the Examiner, "exceeds the standard requirements to such an extent that it provides an exemplary approach", but another recent NDP passed examination on the basis of results from just one extremely minimal questionnaire!

What the Localism Act did *not* do, however, was to strengthen the legislation in terms of engagement in strategic plan-making, despite its ideological roots and despite the previously mentioned government research showing generally poor and unchanged practice.

The legislative framework now in place in England therefore appears to offer good support for effective engagement in plan-making and projects, but the jury is still out on whether that will make a difference in everyday practice.

Legal requirements linked to land use planning

The focus on legislation for engagement in the planning system does not just reflect the main focus of this book; it also reflects the fact that legislation in other closely or loosely related fields is poor or non-existent. Moving to the edges of planning the most important legal constraint is set at a European level on all aspects of environmental policy and projects (hence including land use planning) through what is termed the Aarhus Convention on 'Access to Information, Public Participation in Decision-Making and Access to Justice in Environmental Matters' (European Union, 1998).

One might also expect some requirement for community involvement in other areas linked more directly to planning, for example, Strategic Environmental Assessment (European legislation) and Environmental Impact Assessment and Sustainability Appraisal (UK legislation) but this author is as yet unaware of a single confirmed example of the outcomes of any involvement carrying any weight in final decisions, in fact it is difficult to find any reports on these issues even mentioning community involvement.

Moving further afield, one might imagine that in government literature involvement or consultation (never collaborative working but occasionally engagement) would be a formal requirement in policy and practice for transport, health, education and so forth. Yet, there is either no legislation at all or, putting it bluntly, it is minimal, seen to be gratuitous and therefore ignored by almost all.

Barriers to collaborative working

Given the relatively positive picture painted above and in the examples at the end of this chapter, why is good engagement not an everyday, normal approach? Is the evidence not clear enough? Are there some issues that collaborative working has not addressed successfully, perhaps never can? Are there people, organisations or even societal cultures actually resisting new ways of working because they are seen in some way as a threat?

The answer is probably all of these and more, so it is important for any practitioner to be aware of some of the barriers because they will no doubt be in the air if not explicitly stated, and ideally to be able to produce counter-arguments. Care is needed, however, because over-enthusiastic advocates of collaborative planning, for whom its benefits seem gloriously self-evident, may not be the best people to get across the key messages because they may be blissfully unaware of the arguments and motivations of the sceptics. What follows is a number of key barriers with, in each case, possible responses.

Evidence generally: Is the evidence for any benefits as convincing as it might or should be? It is very difficult to prove cost savings (Involve, 2005), time savings, better outcomes and so forth, and also the value of indirect benefits such as growing confidence in the planning system. Yet, evidence alone has never been the decisive factor in whether or not to adopt some new approach so the next issues are probably more important.

There is no easy answer to this, although the momentum is in favour of more and better engagement simply as a result of people experiencing it working. Being able to quote examples (as later) can help, although it is always better to get a councillor, planner, developer or resident who has experienced it first hand to say it works because people are more likely to listen to someone like them than to an (often overkeen) advocate.

Evidence of benefits: Some systems make it difficult to highlight evidence of benefits, if it is hidden in other indicators for example,[6] or if established indictors show 'disbenefits' to balance against any benefits. A classic example of the latter is about the speed of decision-making on planning applications. It has never been shown that speed of determination bears any relationship to the quality of the eventual development, so any evidence about a slightly longer decision-making process generating better results would be irrelevant and would secure no rewards in the current scoring system.

The answer is to avoid playing this game, focusing instead on all the other benefits that can come from good practice: Better results, saved money, changed attitudes, improved trust etc.

Careers: Engagement work is often something that young planners say they would wish to do but are also cautious about because their profession does not as yet really value that area of work; one is unlikely to be promoted to the next planning job as a result of running even the best engagement programme. Similarly, engagement experience is not greatly valued in career terms for architects and most other environmental professionals.

The answer lies in the argument outlined earlier, that good engagement produces better results, better plans, better designs and it is that which should be promoted first, the reason behind it – engagement – can wait till one is in the next job.

Professionalism: Most professions are resistant to changes that seem to challenge or erode the distinctive skills that they claim to offer. Because they are the experts, it follows that others are not, so finding out and valuing what others might contribute is still seen by some as a weakness rather than a strength. Chillingly, it is still the norm that architectural students in the UK are taught nothing about how to communicate with, understand or work with their clients (direct or indirect). That is less true for planners but, even since the Localism Act, their initial professional training covers little or nothing about engagement.[7]

There will be little progress until initial professional training is better, so the only answer is simple demonstration: Showing the planner, architect or whoever that working with others adds to, not erodes, their professional status.

The media: The media are a different and important barrier. Putting it rather stereotypically, they (or is that we?) love bad news, conflict, disagreement etc. and the last thing they would usually want to report on is agreement: Why let the facts get in the way of a good story? They are also highly opportunistic, so staying with a long-term process is anathema to them.

Hard though it can be, the answer is to do all possible to bring the media into any engagement process from the start, either just by way of information or even as potentially very valuable participants, i.e. to manage the relationship proactively. Feed them useful, topical, interesting material and they will use it, and avoid creating an information vacuum because they or others will fill it, probably negatively.

Developers: The standard point made about developers is that their priorities (in order to achieve the necessary profit) are speed and certainty. This goes some way to explaining their caution about engagement because their experience is that it slows things down and generates choices they cannot deliver. That is, of course based on poor practice either by their planning consultants (who have no training in engagement) or their PR consultants (whose agenda is simply to deliver what the client has already agreed).

Developers watch each other carefully so, if one of them sees a competitor jumping ahead through properly managed engagement, they start to get interested. Good examples, case studies and access to independent and skilled engagement practitioners are all part of a slow shift in attitudes and approaches for developers.

Elected Representatives: As one councillor said on a recent training course run by this author: "I was democratically elected so I don't need to talk to my constituents again until the next election"! This is an even stronger demonstration of the 'I am the expert' argument than described above for professionals, if at least with the justification that elected representatives exist to take decisions on behalf of the community that elected them. Most do, however, rely on being voted in again and so any crude posturing (covered by the media of course) about efficient use of the public's money, stopping a poor development or being photographed opening a nice new park can be of value, whereas supporting a badly needed housing scheme in their ward or constituency can lose them votes, or is judged likely to do so. As with professionals, there is a clear career ladder that does not value collaborative processes for which nobody is solely responsible and there are tensions between supporting things that are overall, or party, priorities and taking possibly difficult messages back to one's constituents.

The first answer is to draw in any elected representatives from day one, in particular in helping to design an engagement process. This helps to focus on the principled approach and on strategic rather than purely parochial issues. That helps them to avoid being seen to be working against their constituents' wishes because it is then the process that leads to the outcome, not just them. Then, of course, if a process is successful, they can claim at least some responsibility for it! They can and should also be used as key links or intermediaries out to the community: 'Community champions' as some have called it.

Community and voluntary groups: Though community groups may be genuinely representative, they too can be as protectionist as councillors about being the sole voice of their community and resent any attempt to (as some express it) go past them to seek views directly from others in that community. And, of course, there are others who cannot ever be described as being representative of their community because they are self-selecting and self-serving cliques.[8] This can also be the case with some voluntary groups (or NGOs), although they are more likely to defend their right to be the sole voice for the elderly, disabled, pedestrians etc. and just as often believe that their group's concerns should be paramount.

For community and voluntary groups it is important to not be seen to be trying to remove or bypass them, focus instead on working with them as partners to use their special knowledge and skills. And this is where collaborative working holds an ace. Instead of a group having to argue their special rights with you as process manager, in collaborative working they have to argue with all the others like them, and that is far more challenging.

The public: It may seem odd to be including 'the public' as a barrier but they are not the barrier themselves; it is people's circumstances and past experiences that can limit their engagement or make it sometimes less than constructive. For example:

- The majority of people may not have even a few minutes of time to get involved; they have families, jobs, leisure pursuits … lives!
- Most are unaware of the complexities, jargon and procedures of planning so they can miss out on things that will, later on, affect them directly (which can cause the classic 'nobody told me' frustration).
- There are also issues around educational levels, gender, class and culture that can limit awareness, ability and confidence to engage.
- By and large all they have ever experienced or been offered has been some form of tokenistic consultation, so how can they judge that any next process will be better?
- From an almost opposite direction, there can also be too many opportunities or invitations to engage and this can lead to what some call 'consultation fatigue' (see Chapter 8).
- Finally, there are some members of the public who are well aware of a process but who choose quite deliberately to stay out of it, perhaps to avoid

facing any form of dialogue with others or perhaps to be able to stand up at the end and say 'I told you so'.

Breaking barriers such as these is a major challenge because it is almost impossible to convince people at the outset that what is being tried this time really is different to last time, but the following can help:

- Being clear about the main principles.
- Bringing in key local people to help to design any process.
- Being demonstrably inclusive and providing regular and prompt feedback to show progress, in particular that people have been listened to.
- Providing several different opportunities to get involved.

And then, if a process is genuinely open and inclusive, there is anecdotal evidence that more people will join in and come along the next time and the time after and that they will value a process even if their particular ideas or aspirations do not form part of the final plan or project. (Chapter 8 highlights some further barriers to progress in terms of making good engagement practice the norm within large organisations, especially local government.)

Does it work?

There is now a long list of projects which, by using more collaborative approaches, have succeeded in securing support for often quite radical solutions and have managed this surprisingly quickly and cheaply. As it is impossible to operate a fully scientific comparison of costs, styles, outcomes etc. (one can hardly try two approaches on the same project with the same people) one has to rely heavily on anecdotal evidence, participants' comments, client reports, media coverage and so forth, to make any proper judgements.

One can, however, compare examples using traditional approaches (Decide–Announce–Defend) with those using collaborative approaches (Engage–Deliberate–Decide). This can either be shown by comparing examples from very similar contexts or, perhaps more powerfully, by considering those where a DAD approach has failed and an EDD approach has then been used successfully.

The examples that follow illustrate that collaborative working and properly managed forms of engagement, even simple consultation, can be effective. Those in Table 3 are short cameos: The column to the left gives an example of a project or plan developed using a DAD approach and the column to the right describes the same or similar example using, successfully, an EDD approach.

Table 3 Examples of DAD and EDD

Decide–Announce–Defend example	Engage–Deliberate–Decide example
Across the UK, many local planning authorities spent upwards of £500,000 each on their statutory plan Inquiries, mostly to pay for barristers. It was common for key policies to change as a result of this lengthy process.	For Kennet District Council, a collaborative process was used, involving the potential key objectors from the outset and reaching agreement on many aspects. The result was a dramatic reduction in the number of objections and significant reduction in the cost and time of the next Inquiry.
In Sheringham, a supermarket company tried for several years to drive a proposal through against the wishes of the community, the planning authority and the Council as a whole, all without success and at considerable cost to all involved.	In East Hampshire, a different supermarket company and the local planning authority shared the costs of an engagement process that so speeded progress towards an agreed solution and an acceptable application that the company saved many months' interest charges.
In Down Ampney in Gloucestershire, a small community (of barely 350 people) raised the largest objection to their area's Local Plan because it proposed 26 new houses for their village.	Later work on a Village Design Statement with the same village community led on to the collaborative production of an agreed plan for the village that included an additional 42 houses.
In Combe Down in Bath, an area of some 500 houses, seventeenth-century stone mines close below the surface threatened extremely damaging subsidence. The authority attempted to impose a solution to infill the mines and generated extremely vigorous protest and resistance.	A new team was brought in, rebuilt confidence in all parties, brought them together and helped to generate a new, multi-party group and an agreed way forward. The plans secured government funding and infilling is now complete.
Bristol City Council was aiming to improve Hengrove Park by selling some of the land for housing and using the money for the improvements, a principle accepted (reluctantly) by the local community. External consultants produced a masterplan without any consultation. The outcome was widespread objections to the plans, which were scrapped, after several hundred thousands of pounds had been spent.	Following a year in which new engagement consultants put the process and relationships back on track, the next set of proposals by a new team of technical consultants was developed through a wide-ranging and continuous involvement process. It went through difficult times but when the planning application was made there were, according to the case officer, "staggeringly few" objections.

Table 3 Examples of DAD and EDD *continued*

Decide–Announce–Defend example	Engage–Deliberate–Decide example
The Tate Gallery in St. Ives wished to expand its art gallery premises in a highly valued environmental setting with a strong and active community. Although they undertook some involvement work, the rather radical designs were sprung on local people. This generated a significant negative reaction which included the formation of a new protest group.	To recover from this position the Tate set up a new local 'Steering Group' reflecting many different views and including a representative of the new protest group. With help from a process manager, the Steering Group designed a new consultation process, arranged a variety of local events and, by the end of a year, had "totally transformed the relationship between the gallery and the town".
North West Water knew that an area needed a new sewage works, in fact local people were calling for one. The manager approached a Parish Council suggesting a site but was rejected. He then did this another 14 times, always with the same negative response.	A new manager brought together representatives from all 15 Parish Councils, people from environmental groups and technical experts. After a series of consensus building meetings a preferred site emerged, agreed even by the recipient parish. As no Inquiry was needed the money retained for that was used to build the works with an extremely high standard of landscaping and to forthcoming (higher) European standards.
The States of Guernsey had tried twice over ten years to get a Waste Strategy approved. On each occasion the plans had been largely finalised before anybody had an opportunity to be involved, so reaction was dramatic and, despite large sums of time and money being spent, the proposals were scrapped.	Third time round the authority decided to try a different, more engaging and collaborative approach. Mainly through a series of very hard, full evening workshops (with two groups of almost 80 stakeholders per session) agreement was reached. These new plans were approved by 74 votes to 1, a unique result.

The next example is fuller and uses a different form of comparison, in this case of much the same challenge addressed differently at roughly the same time in two adjacent cities. It needs to be anonymous because there remains (even after 16 years) serious contention about this set of issues in the authority called A.

A Tale of Two Cities

City authorities A and B were faced with almost identical problems. As the story starts, each has in place a central area 'Controlled Parking Zone' (CPZ), the effect of which is to displace commuter parking to the zones immediately around the centre. Such areas are predominantly residential, so in both cities the streets were, at the time, full of parked commuter cars thus limiting easy access to parking for residents. Both authorities then started to consider extending their CPZs into the adjacent residential areas, at which point the stories diverge.

Authority A followed a classic Decide–Announce–Defend route, appointing external consultants to undertake the necessary studies and report to the council with a recommendation for how to proceed. The study concluded that an extension to the CPZ was needed, suggested areas to which that extension should be applied, and outlined elements of the proposed Residents' Parking Zones (RPZ): The number of permits per household or business, costs to households and businesses, visitors' permits, parking restrictions and so forth. This early work was undertaken with no community involvement or consultation, although a subsequent report recommended minimal consultation because the issue was likely to be contentious.

The study results emerged into the public arena through a leaflet sent to all residents and businesses in the proposed RPZ areas (but not the adjacent areas where further knock-on effects would be felt). The leaflet offered the consultants' option, just one, and sought community views. Public reaction was immediate and vigorous, made worse by the fact that the authority was ostensibly seeking views in a consultation yet had already put in a provisional order for the parking meters! There was an almost unanimous negative reaction, especially to the fact that the ideas had reached this stage without any attempt to involve local people or to offer choices. Emergency public meetings were held, new protest groups formed, judicial review was considered by some residents, and the Highways Department was flooded with letters of complaint.

The net result was a rejection of the scheme in all its aspects. Although the residents of one area suffering particularly badly from commuter parking voted by a reasonable majority to support it, the only choice on offer was for all areas or none, so people in that one area were stymied. Considerable staff resource (supposedly saved by using consultants) had been attributed to doing no more than explaining and defending what had been done; no clear way forward yet existed. The councillors seriously discussed the possibility of shelving the ideas until (as they imagined would happen) things became so bad that local residents would start to request the originally suggested scheme.

The decision to try again led to another supposed consultation via leaflets, and this time sent just to specific sub-areas where the comments had been mixed. This excluded the sub-area where people had been in favour because it had now been decided (with no further involvement) that a RPZ would be put in place there. Again, no mention was made of the possible knock-on effects from the one new scheme, neither was there any further consultation at all with those in the remaining sub-areas where no scheme was now proposed.

Authority B took an Engage–Deliberate–Decide approach. A public announcement that an extended RPZ was about to be considered caused a rush of comments and reactions, many of these as negative as those received, later in the process, in authority A (one community even appointed their own consultants). The authority managed the overall and technical processes in-house but also, most importantly, responded to the initial flurry of concern by committing to thorough engagement, appointing consultants to design and run the process (a team led by this author).

Round one of the engagement involved open, neighbourhood meetings. As well as general invitations, specific invitations went to all known local groups and organisations (such as schools) and to groups in the surrounding areas likely to suffer knock-on effects. Though the boundaries of the neighbourhoods were sensible at the outset, it was made clear that any later boundaries used for any proposals would reflect reaction from the opening meetings.

Nearly 100 people attended each of the 12 local meetings. After initial caution about being invited for their views without any scheme to react to, people became positive about being able to be heard early enough to make a difference. The meetings were mainly interactive, with everybody contributing to generating long lists of parking and parking-related issues, and key principles for a positive future. There was a short presentation from officers about the scope and limits of any RPZ scheme and an opportunity for questions, notably 'how do we know you mean it when you say you will listen to us?'. People were also asked for their ideas on how best to come back to them (and others not present) in a second round of engagement, when proposals were ready.

Following analysis of meeting results, the broad outline of some proposals began to emerge. The next stage of collaborative working was a stakeholder meeting with representatives of around 70 local groups and organisations and including elected members. This event started to focus in on reaching agreement on the details of possible solutions, keeping choices open until the last possible moment, and on developing a clear pattern of appropriate involvement at the options stage.

Following this workshop, a carefully targeted consultation process took place around several possible options about where to use a scheme, permit numbers, costs etc. This was done with leaflets, local meetings and small, very local exhibitions. As suggested by the consultation results, the end result was a scheme applied to some but not all sub-areas, with varying regimes in each neighbourhood. This secured clear community support.

The simple comparison above gives the lie to the old familiar claim that consultation wastes time and money. Far from it. The properly managed scheme (using the EDD approach) was implemented successfully from start to finish in one year. The future for the other (using a DAD approach) is still uncertain over 16 years later!

The 'rolling' example

This is now the place to introduce the main example to be used at several stages to illustrate the application of the key points from each section of the book. This example is offered 'warts and all'. It was not in any way perfect, some things were not done and some failed but, overall, it helped to deliver a widely agreed plan that then led, very quickly, to some practical projects. Once again, because of ongoing issues, it must remain anonymous. In each case the part of the story relevant to the section in the book is told in the shaded box as below:

> The example is about a 'Regeneration Framework' to revitalise a run down, medium-sized town, here called **Waterton**. More about the context and content of this plan will be described at the end of the next chapter. In terms of the some of the issues raised in this chapter:
>
> - **Levels:** Various factors (timescale, budget, scope, attitudes at the time and so forth) meant that the work undertaken was not fully collaborative. Some of what took place could be termed engagement or involvement and there was also a lot of basic information-giving and consultation.
> - **Benefits and outcomes:** There was a strong push towards achieving a high level of community and stakeholder support in order to move quickly onto practical projects. There was also a wish to restore confidence and generate greater engagement from a depressed community; in some neighbourhoods worryingly depressed, even antagonistic.
> - **Principles:** There was a clear overall process, agreed at the outset through discussion. The scope of the plan was clear and the wide range of methods used was intended to be very inclusive. There was a heavy emphasis on mutual education and exchange and on building common ground, ensured by offering a wide range of options. Feedback and reporting were regular and thorough. However, there was no early commitment to abide by the outcomes and (probably connected) it could not be said that all the key decisions were made by consensus.
> - **Legal requirements:** A Regeneration Framework is not one of the main statutory plans; in fact, it bridges between plan-making and project development. However, the intention was to have the final Framework formally adopted by the authority. In that context all the national requirements for community involvement at the time and the requirements in the local Statement of Community Involvement had to be met. However, the Statement was very general and did not include any clear or testable principles, and what was done was beyond anything that had been done before.

> • **Barriers:** Though not a deliberate barrier, the very low level of engagement to date in all the authority's practices (not just planning) made it difficult to move far towards a higher standard. Some councillors were sceptical about community involvement, though others were keen to build better practice. Some professionals, both within the authority and on the technical team, were also sceptical, although the team leaders were fully in support. Securing an appropriate budget therefore proved to be quite demanding as nobody locally had any experience on which to justify the commitment of what seemed to be a large amount of resources. All this background generated considerable caution in the wider community; they had been asked their views before but had never been listened to. This was exacerbated by a notoriously anti-authority local press and media.

Key messages

- Although no 'level' is better than another, all are needed and the aim should be to ensure that every process includes some aspect of face-to-face working, dialogue, deliberation, work in-depth and collaborative working.
- No project is ever likely to deliver all the positive benefits and outcomes described in this chapter but it is only by thinking hard about what might be possible that any substantial benefits will emerge.
- The same is true of the good practice principles; no project will ever score a clean 10/10 against all of them but all should be considered at the outset.
- The fundamental shift that will move the design of any process towards being collaborative (if only in part) is to get away from Decide–Announce–Defend (DAD) and introduce Engage–Deliberate–Decide (EDD).
- Care is needed to ensure compliance with any legal requirements about consultation or involvement (none formally require collaborative working) but good practice as per the principles should always ensure this.
- Be attentive to the barriers, problems, challenges that can limit the effectiveness of anything you do and address them robustly and creatively.
- There are now many examples to use that prove the advantages of EDD over DAD.

Notes

1   Such lists are picked up again in Chapter 5.
2   For example, from the International Institute for Public Participation. Go to: www.iap2.org
3   Thanks to Allen Hickling, original reference unknown.

4   Within the devolved UK system, Scotland and Northern Ireland have their own, different, regulations.
5   Most commentators, this author included, regard the use of referenda as seriously retrograde because they inevitably devalue front-loaded engagement.
6   For many years energy costs for schools were wrapped up with other unrelated costs, making it difficult to even know how much was spent and therefore how much might be saved.
7   With apologies to places where any of this does in fact happen.
8   With the introduction of Neighbourhood Development Plans there have already been examples of two or three groups claiming very vigorously that they and only they speak on behalf of their community.

References

Arnstein, S. (1969) 'A Ladder of Citizen Participation', *Journal of the American Institute of Planners*, Vol. 35, (No. 4, July), pp. 216–224.

Creighton, J. (1992) *Involving Citizens in Community Decision Making*. Washington D.C: Program for Community Problem Solving, National League of Cities.

European Union (1998) *Convention On Access To Information, Public Participation in Decision-Making and Access to Justice in Environmental Matters* (the Aarhus Convention). Available at: http://ec.europa.eu/environment/aarhus/

Fisher, R. and Ury, W. (1981) *Getting to Yes*. London: Penguin.

Great Britain, Department of the Environment (1994) *Community Involvement in Planning and Development*. London: HMSO.

Great Britain, Office of the Deputy Prime Minister (2004) *Planning and Compulsory Purchase Act*. London: Office of the Deputy Prime Minister.

Great Britain, Office of the Deputy Prime Minister (2004/2) *Community Involvement in Planning: The government's objectives*, London: HMSO.

Great Britain, Department of Communities and Local Government (2007) *Lessons Report 3: Participation and Policy Integration in Planning*. London: Department of Communities and Local Government.

Great Britain, Department of Communities and Local Government (2008) *Final Report: Spatial Planning in Practice: Supporting the reform of local planning*. London: Department of Communities and Local Government.

Great Britain, Department of Communities and Local Government (2011) *Localism Act*. London: Department of Communities and Local Government.

Great Britain, Department of Communities and Local Government (2012) *National Planning Policy Framework*. London: Department of Communities and Local Government.

Involve (2005) *The True Costs of Participation*. Available at: http://www.involve.org.uk/blog/2005/11/16/the-true-costs-of-public-participation/

Planning Aid (undated). *Good Practice Guide to Public Engagement*. Planning Aid. Available at: www.rtpi.org.uk/planning-aid/

Skeffington, A. (1969) *People and Planning*. London: HMSO.

3
Getting Ready

Introduction

This chapter is about getting ready, or gearing up, not just for a specific project but also more generally. It starts by considering who is initiating, funding, managing and/or delivering any engagement process because until that is clear there is little point in moving on. This first section deals with 'Putting your own house in order'. Assuming that work is starting on an engagement process, the chapter moves on to things to be checked, set up or put in place before it is possible to produce a process design, certainly before actually starting any engagement. This section is titled 'Beyond the presenting problem'. Some of this is practical, some about establishing a very thorough picture of the specific issue and some about the important and therefore separate issue of identifying potential stakeholders or consultees, so the next section is titled 'First find your stakeholders'. The final section, possible only of relevance once other aspects are clear, is about setting some 'Objectives of engagement'.

The caution made earlier, about things not always being sequential, warrants repetition here. Conclusions about consultees can suggest a need for different skills, for example, graphic rather than literate. Any initial process design has practical limitations which might change who to engage or the scope of the project. And this does not just need to happen at the start; issues and scope can change as work proceeds, requiring additional consultees, new events, different information, even amended objectives. Challenging though this can be, it lies at the core of what makes engagement work so different and potentially positive. It is about being as open and flexible as possible all the way through until final decisions can be made.

Putting your own house in order

Very basically: Who are 'you', what is 'your own house' and on what basis is a choice being made to introduce or support engagement processes, to design or deliver them? The previous chapter introduced one great unspoken of any form of engagement; the assumption that a decision is made by somebody in some sort of power to initiate and deliver a process. But somebody has to do this. The other important unspoken point is about who actually designs, leads, delivers and reports on any process. Such questions are often discussed amongst engagement practitioners but rarely amongst clients and commissioners. Yet both points are absolutely crucial if any process is to attain appropriate legitimacy at all stages.

Clients and commissioners

It is, and probably always will be, central government, government agencies, local councils, companies (especially developers) and occasionally NGOs who are the prime initiators of policies, plans and projects, and hence of engagement. The public can be up in arms and calling for action but that action will most often be taken by one or other of those just listed.

There are, nevertheless, some important opportunities for communities to also initiate certain things, for example, to produce a Design Statement or to create a new local open space. In fact, recent legislation by government has encouraged and supported the initiation and leadership of certain statutory plans at local level, notably Neighbourhood Development Plans, and of certain development projects, notably a Community Right to Build (for example, for social housing). As a matter of principle, whoever leads such projects at community level needs to consult or engage just as widely, ask the same questions and address the same challenges as a local authority or developer.

Given the core principle of collaborative planning that whoever initiates should not inappropriately determine a process and its outcomes, the question becomes one of how an authority, developer or community might best address this. There are three key ways:

1 Adopt the principle of working with others to agree a process. Though the outcome may not be an entirely different process (though details are always enriched with local knowledge), that is not the point. The point is about mutuality, transparency and trust because any process suggested by 'them' is likely to be distrusted precisely because it was suggested by 'them'.
2 Patently abide by all the other principles listed in Chapter 2 and it will be almost impossible for any commissioner to rig the process and distort the outcomes.
3 Introduce some form of separate process manager, deliverer and/or facilitator seen by all to be acting with an appropriate degree of independence.

> In this example, the commissioners/clients were a local authority and a government agency. Neither had great experience of engagement or commitment to it beyond that in the authority's Statement of Community Involvement. However, those managing the process (see next box) were able to agree the process with the wide-ranging project Steering Group, secure support for most of the principles and get recognition of the value of having a separate Consultation Manager.

Managers, deliverers, facilitators

Suggesting the use of someone '*seen by all to be acting with an appropriate degree of independence*' begs several questions:

1. Who are these process managers, deliverers and facilitators?
2. How can they be found?
3. How do we know they are any good?
4. How can we convince everybody that they are acting with '*an appropriate degree of independence*'?
5. Is '*appropriate*' good enough?

The first three questions are unavoidably first because, if such people do not exist or have appropriate skills, then issues of independence become irrelevant. And there is as yet only half an answer. While there are a few courses for trainee planners about the history, theory and key themes of consultation, this author is not aware of any higher education course in engagement, process design or facilitation that addresses practical issues and skills. There is some occasional and fragmentary training at mid-career level but it is entirely optional. If one can find it, that may be a two day course on key skills for facilitation or process design (see Appendix 4). Some of this is linked to forms of accreditation, notably through the International Association of Public Participation[1] (IAP2). To know if someone is any good, the following IAP2 standards are a good start.

- **Purpose**. We support public participation as a process to make better decisions that incorporate the interests and concerns of all affected stakeholders and meet the needs of the decision-making body.
- **Role of practitioner**. We will enhance the public's participation in the decision-making process and assist decision-makers in being responsive to the public's concerns and suggestions.

- **Trust**. We will undertake and encourage actions that build trust and credibility for the process among all the participants.
- **Defining the public's role**. We will carefully consider and accurately portray the public's role in the decision-making process.
- **Openness**. We will encourage the disclosure of all information relevant to the public's understanding and evaluation of a decision.
- **Access to the process**. We will ensure that stakeholders have fair and equal access to the public participation process and the opportunity to influence decisions.
- **Respect for communities**. We will avoid strategies that risk polarizing community interests or that appear to 'divide and conquer'.
- **Advocacy**. We will advocate for the public participation process and will not advocate for interest, party, or project outcome.
- **Commitments**. We ensure that all commitments made to the public, including those by the decision-maker, are made in good faith.
- **Support of the practice**. We will mentor new practitioners in the field and educate decision-makers and the public about the value and use of public participation.

Anybody stating up front and clearly demonstrating such standards is likely to be regarded as appropriate (though this is by no means certain), be they from *outside* a commissioning body or from *inside*. However, this suggests further points to consider about who might best manage and lead.

Someone from outside: External experts usually offer a greater depth and range of skill and experience but this has to be balanced with less knowledge of local contexts. They are usually assertive about their independence, although communicating that to others beyond the process commissioner is always a challenge and can often generate the question that 'if they are paying you how can you be independent?'. This is almost answerable at the outset; it can really only be demonstrated once a process is underway and people see good practice for themselves.

Someone from inside: Even large organisations rarely employ people whose sole role is to design, manage and facilitate engagement. Some members of staff may do such work occasionally but it will not be their main role, so they are unlikely to have the skill and experience that those outside of the organisation may have. Internal staff will probably, however, know the local context well, although they may be, or be judged by the community to be, too close to some of the people, issues, patterns and habits of their employing organisation to be seen as appropriately independent. Beyond that, there is no fundamental reason why internal staff should not take a full lead role or be present to support external people.

The final point to highlight about process managers and facilitators is that not all events are critical. Some are extremely critical and can occur in situations of high conflict. For those, a highly skilled and external person will almost certainly be needed. However, some events are less critical, more relaxed and less intense; for these, an internal staff member can be entirely appropriate. In general, any choice need not be either/or. The outsider may only be needed at the start and perhaps the end, the critical stages, and internal staff can manage much of the other work. Most importantly, whoever the person is, it cannot be assumed that independence is accepted; it needs to be worked at and earned.

> The external engagement team for the project was appointed on a competitive tender basis. Although 'consultation' was mentioned in the brief to the competing teams, only the appointed team included an explicit engagement expert. This expert had worked with the lead team before so some of that team (not all) were broadly in tune with engagement practice. In delivering the work, some activities and events were run with internal staff from the client bodies to share the load and help to build their expertise.

Funding

Unfortunately, there is very little that can usefully be said at present about how much any process might cost, for several reasons:

- Coherent record-keeping of costs does not happen. Ask a local authority how much they spend on consultation in a year and the response will be a blank stare (see Chapter 8).
- Every project really is different, so the appropriate resource can vary significantly.
- Almost all engagement projects access forms of free time or help in kind (for example, a free venue). The effect of this on the necessary resources can be considerable.
- In most cases members of local communities also give their time for free by attending events or, for example, undertaking local surveys. For a Neighbourhood Plan this could be thousands of person hours,[2] worth a huge amount (though one must be cautious about costing voluntary time).
- Most engagement events require attendance by technical team members and their time must also be accounted for.
- Rates for external consultants vary dramatically.

In addition, and in line with good capacity building principles, some consultants will try to spread the load, transfer skills and bring on local people to undertake some of the work themselves. Such on-the-job training may add to the cost on a project but it pays off enormously over time.

If a process is genuinely collaborative it will cost (in all senses) more than one that is solely about information-giving. A sub-regional Waste Strategy will cost far more than for a Village Design Statement. If the process involves people and/or organisations where a lot of such work has been done before (so skill and experience already exist) it will be cheaper than a process where it is all novel.

Very importantly, because engagement is still the exception, most commissioners draft technical budgets without any allowance for engagement and then have to find a way of adding it on later, usually by making cuts elsewhere. Simply by ensuring that at least some sensible allowance for engagement is built into budgets from the very outset, funding is far more likely to be available. And keeping records over time will help to ensure that a sensible 'stab' (it may always be a stab, this is an art not a science) is made next time, and even more sensibly the time after.

Information

For one major international environmental conflict,[3] the informal evaluation suggested that around 80 per cent of the conflict was resolved once people had shared and agreed the key information. Fully shared and agreed information is the bedrock of any successful engagement process, emphasising again the danger of the ladder analogy where information is seen as no more than the poor relation. Though information alone cannot resolve anything, it helps to clear the 'fog' and enable effort to focus on whatever genuinely separates people.

Different people will often hold different information on the same topic. Residents might wave a bulletin that describes the potential effects of an increase in local traffic, the authority might wave theirs suggesting different effects and, of course, a developer might have their own version, different again. There are three things at play here. First, is the actual information itself. Second, the fact that each party waves information at the other but fails to share it. Third, one party may not agree with some information solely because they do not trust who produced it. Put residents, authority and developer together and it might not take long to agree on certain points. The rest may not be resolved but the areas of false difference – the 'fog' – would have been cleared.

Information also comes in various forms. Technical experts usually rely on objective, quantitative information about aspects that can be measured and managed. Others, notably those in local communities, may value more subjective and qualitative, often extremely localised information. Neither is automatically right, both have value and both are needed.

Information can also produce challenges for all parties if it is about the relative value of conflicting aspirations. For example, cars travelling at 20 mph cause more pollution but fewer serious accidents than cars travelling at 50 mph; driving faster generates less air pollution but causes more serious accidents. This is why collaborative approaches are so important because it is only through some level of dialogue or deliberation between parties that genuine resolution can emerge, for example, between accidents and pollution.

Communications media

Even shared and agreed information is of little value if it is not disseminated widely and used properly. Just as different people respond to different methods of consultation, so the same is true of means of communication. In fact, effective communication is about 'belt and braces', using several methods just in case.

The apparent ease and speed of modern forms of electronic communication has obvious value but there is a flip side. Part of that flip side is that some, especially those such as the elderly, homeless, and travellers, may not have access to electronic communication or be able to use it fully. The other part of the flip side is about the speed itself. Dialogue and deliberation, central to collaborative working, are about coming together, seeing each others' faces and behaviours, building up trust, exploring and developing issues, ideas and solutions in a gently cumulative and often reflective manner. In that sense, stopping a workshop for people to talk over a cup of tea becomes an integral part of the process. Much of the chatter through electronic modes is short and sharp and about making statements (often impossible to attribute) which then generate counter-statements and so on, to no real effect. The speed and penetration of electronic communications has enormous benefits but it needs to be used with care if one wishes to move beyond the type of consultation that simply generates long lists of one-off comments that are then resolved by a technician, again in that (still well-used) darkened room.

All forms of communication therefore need to be considered: leaflets, local newsletters, newspapers, radio, websites, Twitter, Facebook, posters, reports in libraries and so forth. Crucially, it also means that some of these methods must be two-way; they need to enable people to respond and feed in, not just receive.

This brings with it an awkward issue. This sort of activity is often thought of as public relations (PR), yet this can be no more than what was described in one project as "just telling and selling", in other words clearly partial advocacy for the client or commissioner and their particular solution, certainly not independent engagement. However, that does not mean that good PR approaches and skills have no place in engagement processes. They do, because process designers and facilitators do not usually offer skills for drafting leaflets, setting up websites and so forth. Someone with good public relations skills can be essential to ensure that leaflets and other media attract wide interest and engage people. The absolutely key point is that any public relations input needs to be one part of an overall engagement process, not the other way round and not managed separately.

- **Funding:** The lead team's bid to secure this project included a sum for consultation as requested, so this was built in from the start. In fact, this attracted attention from the clients post-appointment such that further activities were added by agreement (following the process design workshop with the Steering Group) and extra budget was attributed.
- **Information:** The available information proved to be spread extremely widely amongst statutory, private, voluntary and community and was in conflict in some quite significant ways, notably in relation to traffic (data was out of date and disputed). This was tackled in an opening stakeholder workshop in particular but needed to be addressed on several later occasions.
- **Communications media:** Even though this project took place fairly recently, it was a major challenge to persuade the clients to set up a project website. They insisted that it be on the main local authority site, which inevitably devalued it in the eyes of some local people. The work also took place before social media (Twitter etc.) had become standard, so considerable effort was put into engaging the editors of the two main local newspapers, with general success in that they shifted (if not always) from their established anti-local authority stance.

Beyond the presenting problem

Consider this short meeting between doctor and patient:

> *Patient*: 'Doctor, I've had a rash under my arm for a month or so. Can you give me some cream to put on?' The doctor takes a quick look under the patient's arms and writes out a prescription for some cream.

This was clearly too quick and superficial, so a more appropriate alternative conversation might be:

> *Patient*: Doctor, I've had a rash under my arm for a month or so. Can you give me some cream to put on?

Doctor (having looked under the patient's arms): Have you been under any particular stress recently?
Patient: No.
Doctor: Have you bought any new shirts or vests?
Patient: No.
Doctor: Have you bought any new deodorants or soaps?
Patient: No.
Doctor: Have you changed your diet?
Patient: No.
Doctor: Have you changed whatever you use in your washing machine?
Patient: No ... oh, hold on, yes, I did change the liquid about two months ago. Someone suggested I ought to try one of those biological ones.
Doctor (after a couple more checks under the arms): You don't need any cream, but you do need to go back, right now, to your previous clothes washing liquid.

The outcomes of these two scenarios are startlingly different, in fact, the first would probably have seen the patient back again soon because the doctor had not asked the key questions to get past what is often termed the 'presenting problem'. Now for an example more directly from the territory covered by this book:

- As part of engagement work on a Management Plan for the Blackdown Hills Area of Outstanding Natural Beauty this author convened a Topic Group on Traffic and Transport.
- The group included local community people, a local councillor, a local authority footpaths officer and two highways engineers.
- At the first meeting all the locals said the same thing, in effect, that "something has got to be done about all these outsiders, non-locals, who go speeding like mad through our villages".
- With that problem in mind (call it A) the engineers were to come up with a solution (call it X).
- The author let this discussion run a little and the issue emerged again at the second meeting.
- After some more discussion, one local person sat back in his chair and said: "I have to be really honest, it's not the outsiders who speed through our villages, it's us, the people who live here".
- The other group members nodded rather sheepishly, rather reluctantly and very noticeably, but all agreed.
- This was now not problem A (outsiders) but problem B (locals). So the engineers' solution X would probably not work and a different solution was needed.

Only very rarely is the initial presenting problem the real problem, the only problem or even the most important one. Digging down, looking round the edges or over the top, prodding and prompting about whatever issue, problem, opportunity, plan or project is the subject of an engagement process is absolutely crucial to eventual success. Furthermore, effort put in to getting to the genuine problem(s) at the outset will rarely provide the full answer because further factors almost always emerge as work proceeds. That is because, at the start, people may not wish to raise key issues at all; they may, for example, raise every possible issue about a new development other than its effect on their house value. Sometimes it is only as work proceeds that people, including the engagement manager, see the relevance of something they had overlooked and it is only once trust has been built that people will open up to what really worries them.

Asking questions, pushing and prompting to get to the key issues almost always gets answers but it can be too easy to get seduced by those answers because some may have nothing to do with the situation at hand; they are the classic 'red herrings'. Separating those issues that are genuinely relevant from those that are unrelated is extremely important and, to be honest, really only comes with experience.

Who to question, how to question

While it is usually inappropriate to rely on the clients or commissioners alone, they are the people to start with. As the professional holders of the issue and usually the lead implementers of any resolution, they are likely to know most about all aspects and are, as will be obvious shortly, the only ones who can answer certain questions. This can also help to clarify, if not determine, who else needs to be spoken to.

The words 'spoken to' are important because it is far better to do any probing or questioning face-to-face. At the same time, to help whoever is being questioned to be ready to discuss something potentially challenging, to reassure them and to use time spent with them efficiently, it is helpful to share with them in advance some of the main questions, perhaps via a short email or telephone call.

Once face-to-face with someone, use any questions in a gentle manner rather than as an interrogation, asking and listening in an understanding and helpful way. Asking them why they think a question is being asked can be especially valuable because it can help the person think about how their answer might possibly influence the upcoming engagement. By discussing in this way, rather than just ticking boxes on a questionnaire, they are implicitly learning about the added value of thorough engagement.

In particularly tense and conflict-ridden situations it is often critical to be sure to speak to *all* key actors, and also to be known to have done so, in a procedure described as 'shuttle diplomacy' (explained more fully in Chapter 4). This may have to happen before people will even be willing to get in a room together.[4]

Questions to ask

The questions that follow are in a generally proven sequence but it is important to feel free to alter the sequence, add or remove questions and move between questions, allowing the conversation to flow. It can also be necessary to revisit an early answer as a result of information gained from later questions and it is important, especially in high conflict situations, to feed back and check any summarising of answers given.

For all the basic questions that follow, there is also some commentary on why that question is important, i.e. what the answers might suggest about the future engagement process, methods or consultees.

Written-up notes of any conversations should be produced and, wherever possible, sent back to each separate person and their correctness agreed. For contentious issues, the overall results may not be shared round with all (and may even be seen only by those managing the engagement) except by agreement by all. For non-contentious issues, sharing common points with all (again by agreement) can begin to remove unnecessary differences of opinion and create a commonly agreed starting point.

The history of the plan or project: Is it a recent idea or long-standing? Has it been tried before? With the same scope or aspirations? How has it been received previously? To what effect? How have any previous attempts been handled? With or without consultation? If some, what sort? How was this received?

This is a crucial first question because only rarely does any plan or project arrive totally fresh and new; there has often been some previous history to it. This can be very direct (about a previous attempt at the same project), indirect (about a similar project elsewhere), recent (just last year) or a long time ago (somebody once quoted a project from 37 years ago). Any of these can create what might be called 'baggage' that people will inevitably bring with them when a new process starts. That baggage may be irrelevant or marginal but it still needs to be recognised or it can have an insidious and damaging effect. There again, some of the history may be positive and that should be built upon.

Other initiatives or consultation under way: Are there any other initiatives under way in the same area? With consultation/engagement? Has anything happened recently or is anything happening nearby that might have a knock-on effect on the project?

Little can be more damaging than to start a good process and then discover that others in the area are working on the same topic or are starting other poor quality initiatives (or even other good ones) in the same place at the same time. It can damage trust by suggesting that 'they' (manager, clients, commissioners) do not have their act together. Check carefully and, if at all possible, try to share or at least avoid obvious conflicts.

The scope for change: What is totally fixed? What is potentially changeable? What is clearly open to consultee input?

This directly reflects one of the principles outlined earlier. It may appear to be a question solely for the project commissioner but the answers can usefully be shared with others because scope is never set in stone. It is also worth challenging (gently) to establish how certain someone is about their answers. They may be expressed with certainty but that may not be entirely the case, and where the answers come from is important because a person may be responding on behalf of someone else, or just making assumptions.

Project timescale: When might things start? Are there stage deadlines along the way? Is there a critical date for completion? Are these all fixed and, if so, who by or what by?

Most commonly, the response to this comes from the project commissioner. They may state a given date or set of dates for agreement or completion of a plan, for example, to present to a Committee or submit a planning application. Too often, as with budgets for engagement, timescales and target dates are set without any consideration of appropriate timescales for engagement. It is not so much that timescales are wrong but that they have emerged for the wrong reason, for example, the technical programme that ends up with consultation over the Christmas/New Year break. Quoted timescales are often also not as rigid as someone thinks, so it is important to press on this because time itself is an aspect of scope. And none of this is necessarily about seeking additional time because good engagement can speed progress.[5]

Resources and their availability: What budget has been allowed? Does this include for time from others? Does it include all the administrative details (venues, mailing, leaflets, travel, catering etc.)? Are there appropriate and available venues? Are there events or activities to 'piggy-back' (e.g. a community newsletter or festival)? Is any free help possible or available (e.g.

leaflet deliverers, informal surveyors, free venues)? Are people with the right skills available?

These questions are best asked of everyone, not just the project commissioner. Having focused earlier on funding, it is important to recognise that cost itself is sometimes not the issue because there can be a remarkable number of things that can be procured for free.

Status of the project: Is it the same priority for the commissioning organisation as for others, especially any local community? Is it seen as hard, intractable, easy, or contentious, as a diversion or a quick win? Who sees it this way and why?

Once again this is not just for the commissioner because other key stakeholders may have quite different opinions. How it is seen can affect resources and commitment, scope and timescale. This can also change over time if, for example, a particular problem emerges (a newly-formed protest group) or other priorities take over (new legislation demanding a shift around of staff for an authority).

Taking final decisions: If the project is a statutory plan, decisions rest mainly with an external Inspector. If it is about a planning application, must the application go through councillors or can it be decided on delegated powers? If it is a private sector project is there some sort of senior management or Board who must approve it? Does it require sign-off by others, for example, statutory consultees or community?

It is important to know how the results of any process will be used, and by whom, in decision-making. That can affect what evidence needs to be produced and in what form it needs to be presented to make a convincing case about using engagement results and not being swayed by last minute pressures. Last minute pressure still comes, far too often, from councillors (the archetypal speech at Committee) which further reinforces the need to draw elected representatives firmly into any engagement process so that they become supporters.

Information available: What (only generally at the outset) is known now and to whom is that information available? Are there some key things on which further information could usefully be collected and made available now, before starting? Who might commission, assemble and disseminate this new information? What status does current information have amongst those likely to be involved?

If little information is available at the outset it can or should affect the project timetable because some information collection may take weeks, even months to complete (some ecological information can only come from surveys done at one specific time of year). Knowing the nature and extent of information also provides clues as to whether it is, or is likely to be, widely accepted or whether real work (and time) must be put in to agreeing it.

Key participants: This is of major importance so is elaborated in the next section. As for the previous topic (Information), all that is needed at this stage in any process is some initial suggestions.

'Snakes and ladders': This is about things of relevance at a different level. In any situation there can be some 'snakes': notorious spoilers, new people/groups, reduced resources, hidden agendas, raised conflict, external decisions and so forth. There can also be some 'ladders': growing confidence, early wins, positive feedback, levering in added resources and so forth. This is a form of risk and benefit analysis[6] but deliberately done on a more subliminal level, aiming to capture things that a formal analysis might miss. Knowing about these 'snakes and ladders', where they are, who they relate to and so forth, can be crucial in designing a process.

A final question needs to be used to seek a reflective summary from the respondent, highlighting what they judge to be the key issues or uncertainties. As suggested, asking such questions will almost always raise issues that the respondent has either forgotten about or put to the back of their mind, or which are perhaps quite new. That is why this question is asked at this stage.

> The main statutory plans for Waterton had taken years and often been overtaken by events, while there was almost a full graveyard of failed projects. Those in the authority blamed this on an unresponsive and negative community. Those in the community blamed it on an unresponsive and negative council! There were in fact, some communities where, despite genuine attempts, severe disadvantage and disaffection remained. The flip side of this was the number of potential sites in the town ripe for regeneration and the scope to do something about some of them relatively quickly. Some aspects were not really 'up for grabs', notably main highway and transport solutions, there were some projects already in the pipeline and difficult to stop and there was a need to link the regeneration framework into the overall district plan.

> The initial timescale seemed reasonable at the time it was created. Although it just missed one key community event that could have been piggy-backed, it managed (after a couple of minor delays) to link into another one. As usual, it was a lack of technical information on traffic that led to a time extension. The basic budget was set at the outset but extra (free) support was secured from the local media (using their websites), local businesses provided venues for events and some community groups undertook some of their own survey work. Links to a major national industry drew in extra engagement budget.
>
> For the local authority involved, this project was seen by some staff members and a few but not all councillors as a key opportunity to improve practice all round. This only happened because of the involvement of the external regeneration agency whose staff had been through several high profile engagement initiatives before. Some influential people in the community, notably business and environmental interests, also saw this as a potentially habit-changing opportunity. The engagement work was therefore in the spotlight from several directions. To balance this high profile, there were real anxieties about where final decisions might rest because a local election was due; the ruling party leaders were anxious that the project might affect voting patterns negatively (for them). At the same time the opposition party was starting to make noises about a waste of public money!
>
> There was plenty of information available but it was often out of date, not specific to Waterton, very one-sided and debatable, and there were huge gaps on key issues such as community facilities and retail. (Consultees and stakeholders will be picked up next.)
>
> In this case the main 'snakes' were also potentially the 'ladders'! The election brought many otherwise hidden, long term issues to the surface (good and bad), the disengaged community which could have been a problem created a very open canvas on which to try things out and the projects in the pipeline, even the less good ones, could be fast-tracked through to look like quick wins for the Framework and hence build confidence in the initiative.

First find your stakeholders

Who might be 'stakeholders' and 'consultees'?

Careful, effective collaborative working is crucially dependent on who the collaborators are and on engaging them appropriately. However, 'collaborator' is not a useful term. 'Consultees' is a familiar term but it carries undertones about processes of minimal involvement rather than genuine engagement. 'Participants' is fairly neutral and translates well into languages other than English and is probably the least worst. In everyday conversation, the term 'stakeholders' also comes across rather uncomfortably but is used here because, once its meaning is explained, it is probably the most useful term.

The term 'stakeholder' is meant broadly, to include all those who have some form of stake or interest in a specific situation, plan or project. While 'consultees' are usually construed as those potentially impacted directly by a plan or project, notably those living in a particular area, stakeholders can have several other interests. They can be:

- Gatekeepers on legal issues or other standards, for example, the Environment Agency as the body which grants licences for waste schemes.
- Eventual decision-makers, notably local authority elected members.

They can also have downstream interests because their involvement, agreement, support, skills, even money could enable or block later delivery, for example, the Police on detailed security issues.

Some stakeholders have interests in all aspects of a situation while others may be concerned about only one, but to them very significant, aspect. Even within a local community, where one might expect people to be concerned about all aspects of a proposal, some may only be interested in traffic or design (or even one particular tree, as happened recently).

The next factor to consider is the word 'local' because it is used a lot (as in 'the local community') but also too often glossed over. In research by this author some years ago in Milton Keynes, one resident, when asked to draw her neighbourhood, drew her house and a few other houses around it. When asked the same question, the woman living next door drew the whole of Milton Keynes! In another example for this author, for a wind power project with inevitable visual impact issues, the requirement was initially to talk to communities living within one mile of the proposed site. However, those communities lived under a steep slope and would never have seen the turbines from their homes. It was the communities living well over a mile away who would have been most affected visually.

There is still, far too often, a knee-jerk assumption that 'the local community' for a housing development, for example, is simply those living near the site. Yet a possible development could generate enough traffic to overload a junction some way away, have a significant impact on a key view or include a community facility that would attract people from many miles around. Equally, people who come into that area, for example, to shop, are important even if they are not strictly local, because their continued use (or not) of the shops could be critical for the future survival of those shops.

Despite this, there is a general correlation between the size of any development and the physical scope of any plan and the community to be engaged. What is more, this should not be affected by local authority or parish boundaries. If expansion is proposed on the edge of a major town, engagement should include some in the major town even if it is in a different local authority area. Traffic movement is a useful example for this because it can create linear patterns of potential involvement, creating impacts that could be felt some way away along a particular route.

Similarly, there is a general correlation between the nature of any development and its potential users, impacts and hence stakeholders. If a swimming pool for a whole city is being proposed in a new residential neighbourhood, the community for that centre includes not just those living in the adjacent neighbourhood but potentially everyone in the city; a significant challenge for any engagement.

Addressing such challenges is exactly where 'stakeholders' can be of most value. Take the swimming pool example. Contacting, even just identifying, perhaps 10,000 swimmers in a city is simply impractical, but many swimmers will be part of some form of swimming club and those managing other city swimming pools will be well aware of issues and concerns amongst swimmers generally. With that in mind, perhaps a dozen people could be identified and contacted to ensure that a reasonable 'swimming voice' is heard.

Yet this is the aspect on which engagement focused on stakeholders is most often criticised because it appears to be so narrow, selective and even exclusive. Such criticism is not entirely wrong if a process focuses solely on stakeholders because there are many people whose views are not in any way represented through a group or organisation because they do not belong to one. Hence the point made already that good processes engage in-breadth (to all) *as well as* in-depth (to a few). There is also a rather blunt point to make in defence of stakeholder approaches. As practical experience has shown, the total membership of those groups represented at a stakeholder event for a project in an area can often be greater than the number who voted for the councillor for that area and certainly far more than one could ever attract to a completely open, public event or persuade to fill in a questionnaire.

And finally, just to complicate matters, an important issue to which there is perhaps no more than a glimmer of an answer. All of the above is about hearing from those already in a place or area or near a potential site. How can one hear from those who might benefit from a future development; people desperate for a home, somewhere to meet, a school for their children who currently have to travel miles to be educated? Their voices are never heard directly but the 'glimmer of an answer' is that good quality social and market research can provide valuable information to help ensure that what they might want is at least considered.

Many such examples can be suggested but the key point is that real thought needs to be given to deciding what is 'local', who is 'the community' and who are 'the stakeholders'; simplistic, knee-jerk responses are not good enough.

Identifying stakeholders

This section comes with a health warning: Lists can be very long indeed. Nevertheless, preparing a list is a key task at a very early stage of engagement (and Chapter 4 shows how one can deal constructively with very long lists and make them manageable). List-making is a task best started when asking questions of key people as in the preceding section. It can also be a task for some sort of small steering or reference group. Whatever approach is taken, various different people should contribute and it is important to be aware that any initial list will almost certainly change as work proceeds.

In terms of basic categories of stakeholders, the most common focus is on groups and organisations within the following basic, and not exclusive, classifications:

- **Sector**: Most usually defined as public, private, voluntary, community.
- **Function/Interest**: For example, user, service provider, regulator, landowner, developer, resident, specialist (e.g. architect, ecologist), decision-maker.
- **Geography**: Immediately adjacent/close to the plan area or project site(s), close by, in the broader neighbourhood area, further afield.
- **Socio-economic**: Income, gender, age, length of time living in area, ethnicity etc.
- **Affect**: Directly affected, indirectly affected, previously affected, not affected.

The four categories within 'sector' warrant further explanation:

- **Public sector**: Organisations funded through the tax payer via central government (for example, Highways Agency and Natural England) and via central and local government (notably local authorities).
- **Private sector**: Wholly funded through private enterprise. This includes everything from multi-national companies to very small businesses. There are some significant categories within this that should not be overlooked to ensure good coverage; for example, retailers, light industry, utilities (gas, electricity and others), communications and engineering.
- **Voluntary sector**: Often referred to as NGOs (non-governmental organisations), these are funded through grants, charitable donations, membership subscriptions and so forth. This includes national and local branches of charities and campaigning groups (for example, the Campaign to Protect Rural England and Age UK) as well as smaller local organisations. Most will have paid staff although they usually also rely heavily on volunteers.

- **Community sector**: This covers a huge and varied medley of community groups, only occasionally with staff. This is everything from an informal local book group, through fishing groups (memberships of which may nevertheless be in the thousands) to amateur football clubs, arts groups and so on (see Bishop and Hoggett, 1986).

Note also the increasing number of social enterprises that bridge private and voluntary sectors, essentially run as companies but usually on some form of cooperative model and 'not for profit'. There are also some cross-sector organisations (for example, Town Centre Management companies and Local Enterprise Partnerships) jointly funded by the public and private sectors and often with some voluntary sector involvement.

Almost all of these in any particular area are likely to have already been identified by a local authority in its Statement of Community Involvement so contact with them is relatively easy, although SCI lists are rarely kept up to date and contact names and details for community groups in particular can change with alarming rapidity (as can their actual continued existence).

It is not, however, anywhere near so easy to contact those who do not belong to, so are not in any way represented by, any of the above. The local Electoral Roll is a start and can be used, but not everybody registers to vote. Lists of those paying Council Tax would get closer but they are not publicly available. Community Associations can often provide lists of local residents if an initiative is focused on a particular small neighbourhood, but again these lists are often incomplete or out of date.

Two final points must be made here. First, the Regulations to support the 2004 Planning Act included a long and rather odd list of statutory consultees for formal plan-making (see Appendix 2), all of which must be checked for authority-led projects, even if some on the list can be then struck off if good arguments for this are put forward. Second, the establishment of any database, its use and potentially sharing it with others, must be managed with serious attention to legal issues of data protection. For example, in the UK legal context, if people are invited to and attend a workshop, all names can be shared and a report can be emailed to everyone using a single list. However, if people sign in at an exhibition solely as individuals, any feedback must *not* include the list of attendees, so 'blind copies' must be used for emailing.

Preparing the list for this project was a time-consuming task. The basic list of statutory consultees existed but was incomplete and out of date. The list that was eventually used became almost twice as long and others frequently had to be added as the scope of the project changed. After considerable hard work, the final list of stakeholders at the outset – which then changed as the project progressed – included, amongst others:

- Waterton Heritage Regeneration Partnership
- Waterton Rotary Club
- Wessex County Councillors
- Waterton District Councillors
- Some Parish Councillors (from nearby villages)
- District and County Council officers
- Waterton YMCA
- Association of Waterton Industrialists
- Waterton Arts Centre
- Wessex Access and Inclusion Network
- Planning consultants (for key landowners)
- Waterton Energy Centre
- Locally located national industries
- Waterton and District Civic Society
- Wessex Health Care Trust
- Sustrans (cycling group)
- West End Traders (from part of Waterton)
- Waterton Job Centre
- Waterton Tenants Voice
- Homes in Wessex (social housing company)
- Chidwell School Council
- Waterton Council of Voluntary Service
- Wessex Children's Aid
- Waterton Coaches
- *Waterton Evening Post* (newspaper)
- Waterton Action Group
- South Area Trades Union Council
- Wessex Arts Education Partnership
- Wessex Police, Fire and Ambulance Services
- Waterton Federation of Small Businesses
- Waterton Chamber of Commerce
- Waterton Town Centre Partnership
- British Waterways
- Wessex Tourist Information Centre
- Churches Together in Waterton
- Age Concern Wessex
- Wessex Sports Partnership

- Home Builders Federation
- Wessex Women's Institute
- Road Haulage Association
- Wessex Racial Equality Council
- Regional Association for Better Transport

The 'hard to reach'

Beyond all those outlined here, there are some who are often described as 'hard to reach'. Statements of Community Involvement are supposed to list those to whom particular attention should be paid. One SCI mentions young people, people with disabilities, members of the black and minority ethnic population, and people who rent rather than own their home. Another mentions elderly people, young people, ethnic minority groups, gypsies and travellers, faith groups, disabled groups and those living in rural areas, particularly the more isolated areas.

There are some anomalies here because, in most areas, there will be at least one group representing elderly people, similarly for disabled people and those from ethnic minorities. Such groups are not therefore hard to reach, but that comment requires qualification. Those who run the groups representing, for example, elderly people may not themselves be elderly but may be (often much younger) paid staff or volunteers. The group may also have no formal members, merely stating that they speak on behalf of elderly people. This is not to denigrate their input or expertise but simply to make clear who might actually come to an event representing the elderly, and it has to be recognised that many individual elderly people *are* hard to reach.

Young people are, however, a genuinely hard to reach group and one commonly recognised as such. Outside of school or college, they are an assertively fragmented, often highly individualised group; in fact, it may even be wrong to even think of them as a group as the last thing some of them want is to be labelled that way. As a result, if they do come together, they often do so in an anti-authority way, if not necessarily as gangs then certainly as informal groups unwilling to join in adult-led (they might say adult-dominated) processes. It is currently difficult to know whether the growth of social media amongst young people has helped or hindered this issue but it does at least offer a point of potential access. One can of course access many young people through schools or colleges, which is good in basic principle but they then become a semi-captive audience and this may influence any contribution they make.

More recent debate about the hard to reach has moved on, in two ways, from simple SCI lists and the assumptions that go with them. First, some prefer the term 'hard to engage'. A good example is people with disabilities. Local authorities have registers of almost all disabled people and there are many groups at which they meet, so they are relatively easy to reach. However, it does not follow that those contacted will ever engage, perhaps because of a lack of self-confidence and a feeling of being undervalued, perhaps because of simple physical or mental ability, i.e. getting to a workshop at night, using social media or even having a computer.

The second query that has arisen is about definitions themselves. For some elderly persons' organisations, 'elderly' starts at age 50, but the main concerns of almost all groups working on behalf of elderly people are with the challenges (finances, frailty, mental competence, mobility) that typically come with ages well beyond 50, even 60. Many 50-year-olds would be insulted to be placed in the same category as those aged 70 or 80, but who then represents them? The answer is nobody, because not only is forming an age-based group not seen as relevant by most of those aged 50 to 60 but, if there *was* a group, they would probably never join it! The same applies to families (especially single parent families) with young children for whom household finances may well be stretched and where time to do anything (for example, participate in a group which might then represent them) is simply not available.

This effectively brings us back to the earlier point about all those who are either members of no group or who have nobody speaking on their behalf. They can be hard to reach (because no contact details exist for them) or hard to engage (because they are known about but are too busy or suspicious). How to break through these apparent barriers has taxed engagement specialists over the years. It is particularly difficult within the general levels of budget attributed to engagement, because it requires considerable effort and special skills, but even allocating considerably larger sums of money seems unlikely to truly succeed. This is a difficult if important issue but no quick and easy advice can be offered here because each group in each context is different. Chapter 5 does however offer some suggestions that have at least been shown to help.

> There were some extremely disadvantaged communities in Waterton. They were identified as a key target in the original documentation about the project, as were young people.
>
> The engagement team made little attempt themselves to access the disadvantaged communities directly, working instead through those who already had a way in to the key communities, for example, community workers, health workers and youth workers. They were able to bring at least a few people together locally and to introduce the engagement and technical teams. Some valuable workshops were run with local people, although numbers dropped off as work proceeded and it proved impossible to persuade any of them to attend any of the project workshops where they could work alongside many others. (Some community workers came but that's not exactly the point.) This was, at best, a limited success.

> The initial approach to young people was through local schools and the college. A number of patently fun and enjoyable events were held with young people from 8 to 18. Some did visioning work ('A Day in the Life of Waterton in 2020'), some solved clues to get around the town, unwittingly providing consultation results while they did so and some trawled the web for images of the sort of spaces and buildings they wanted in the future.
>
> Because they were then persuaded to join in some of the other engagement events outside of school/college, this had several (planned) knock-on benefits. They got excited about what was happening so brought their parents along to events, and the project team were able to excite people with all sorts of images that most of the older, more conventional stakeholders would never have suggested. This culminated in a Design Day when seventy 14–16-year-olds developed 3D designs for several of the identified key regeneration sites in Waterton. This generated further good publicity and one design was so good that it was used, with a bit of extra graphic work by an urban designer, as the frontispiece to the final project report!

Objectives of engagement

Each participant or stakeholder group in any situation will have their own specific, and probably different, *content* objectives or hoped-for outcomes in mind. For example, faced with a development on the edge of a village:

- The residents' main objective might be to retain the village's distinctiveness.
- A landowner in the village, about to retire and wanting a good pension, may be aiming to secure a price for the land that enables a comfortable retirement.
- The key objective for the local authority, especially its councillors, might be to have a sound statutory plan in place before any application comes in.
- For the developer of the housing scheme, the key objective might be to gain a speedy planning permission that maximises financial return.

As with the questions to ask, there may be more hidden objectives behind this, for example:

- The residents might be concerned about design quality only because of a concern for their house values.
- The landowner might wish to carry on living, after retirement, in the same area so might be concerned to avoid damage to long-term local friendships.

- A councillor for the village might wish to propose a particular form of development to maximise the chances of being voted back in at the next election.
- The developer might be looking at other sites in the same area, so is concerned to maintain a positive reputation with the community and councillors.

Once again, by digging down beyond the presenting problem, collaborative approaches can help to highlight the more hidden and perhaps slightly selfish content objectives and bring the more positive, win/win objectives into play; in this case a land value linked to a good pension *and* a good design that does not challenge the emerging plan *and* delivered speedily with local support.

With this in mind, the objectives of the *engagement or the process* can be developed not just for each party, because they are likely to be fairly similar anyway, but jointly. For a village development, a set of engagement objectives might be agreed by all as follows:

- To bring together all relevant parties to raise awareness and understanding of the limits and potential for the ... development.
- In the light of the contributions from all parties (time, information, ideas, willingness to share etc.), and recognising their different as well as complementary objectives, to decide together in an inclusive, open and transparent way, the content, form and design of the development.
- And to do so bearing in mind what is agreed to be technically feasible, acceptable to all, financially viable and environmentally acceptable.

The observant reader will have noticed that this statement almost repeats some of the earlier benefits and principles of good collaborative working. That is no surprise and to some extent the objectives above are self-evident and hence probably agreeable to by all, as here. The key point, however, is that this separates the often conflicting *content* objectives of each party from the *process* objectives, not allowing the former to be lessened or become partial to any one party. Keeping content and process objectives separate is in fact the only way that apparent conflict on content can be resolved. Without this, the engagement objectives, even for supposedly more altruistic bodies such as the local authority, become nothing more than a way of delivering each party's separate content objectives.

Put simply, possible benefits and positive outcomes can be re-expressed as objectives or even targets. In addition, agreeing these at the start is remarkably valuable later because they become the pivot point around which the plan or project can be tested as it evolves, and they form the basis on which later evaluation can be undertaken.

Key messages

- Whoever initiates any collaborative process must show clearly that they are not determining the outcomes and work hard to achieve this.
- Independent, external designers, but especially facilitators, can be best, but there can still be many occasions or events where an internal person can be perfectly appropriate.
- Before actually starting any engagement, it is crucial to put in place the necessary funding, set up a thorough database, establish key information and set up a variety of communication media.
- It is difficult to over-estimate the importance of asking questions to really understand the 'symptoms' before suggesting the 'prescription' of any engagement process.
- It can be time-consuming but essential to have in place a thorough list of all possible consultees and stakeholders, and to accept that this will change as work proceeds.
- In developing a stakeholder list, avoid superficial ideas about 'local community' and work hard to identify and then access and engage those often termed 'hard to reach'.
- Be clear about, formally record and keep referring back to the overall objectives. These can perhaps be thought of as a translation of general benefits into specific targets.

Notes

1   Go to: www.iap2.org. Note that IAP2 also still uses the term 'participation'.
2   For work on one aspect of the strategic plan for Bath and North East Somerset Council, the authority estimated that the community contribution had been an astonishing 384 days!
3   The potential disposal at sea by the Shell company of the Brent Spar oil storage platform.
4   When working on the parking strategy for Stratford-upon-Avon it emerged at the outset that representatives of two key groups had refused to meet together for over ten years to discuss parking.
5   It is this author's experience that technical issues, notably about highways and traffic, are usually the cause of delays, not engagement.
6   It is fundamental in consensus building that any *risk* analysis is also accompanied by a *benefits* analysis.

Reference

Bishop, J. and Hoggett, P. (1986) *Organising Around Enthusiasms*. London: Comedia.

4
Design to Deliver

Introduction

This chapter, along with Chapters 5 and 6, form the core of this book. They are all about engagement actually starting to take shape and then happening. By assembling the issues, constraints and opportunities generated by considering a project's context, history, potential scope for progress and the different participants, one can be ready to start producing a design for the overall process. Overall design is the focus of this first core chapter.

The word 'process' has been used several times already. It is fundamentally important because success is not about clever tricks, one event or method, or instant answers. It demands an evolving 'story' that slowly brings people and issues together towards some level of agreed resolution.

To have a good process always involves a form of design, something far more than just choosing method A to be followed by B, then C and so on. Design is about coherence and integration, seamless sequence, bringing the right people and issues together at the right time in the right way, ensuring that varying types of result can be assembled to develop an argument, managing the move from one stage to the next and finally enabling people to find those elusive but critical win/win solutions. However, rather than design as in architecture (where the outcome is fixed and is built as designed), engagement processes offer, and one must be ready to take up, opportunities to review, learn and adapt as things proceed.

The other key point is that process design is worryingly badly covered in the literature on engagement, consultation and so forth (two exceptions being Driskell, 2002 and Environment Agency, undated). The literature is heavily focused on methods; indeed there are some books and websites that offer just one method as the miracle solution to everything, and some organisations which offer, and in some cases sell, just one type of event as their single 'fix-it-all'.[1] There was even one nationally funded guide, produced following the 2004 planning system changes (but now unavailable), that missed out overall processes altogether in its approach to managing involvement in planning.

Continuing the analogy of healthy diet, methods are the equivalent of ingredients, events the equivalent of recipes, and processes the equivalent of whole menus or meals. Who would be happy with a cookbook that did no more than list ingredients or stated that any of 20 ways to cook cabbage produces the perfect diet? As any good cook would agree, a great recipe cannot make up for poor ingredients (methods) but it is how they are combined that really makes for a great recipe (event) and then how courses are combined into a menu that makes for a great meal (process).

There again, as any teacher of architecture would agree, it is not easy to talk, explain or write about how to design. Equally important is the fact that this and the next two chapters could easily have been written in reverse order! Although this chapter focuses on overall processes, designing those processes depends on good knowledge of what sort of events or activities could be run (as in Chapter 5) and deciding on events depends upon having at one's fingertips a whole repertoire of ways to manage people (as in Chapter 6). All are essential. Very importantly, read this chapter *and* the two that follow *before* moving on; it is only *after* all three that the whole picture becomes clear.

Getting to the starting blocks

As stated already, few if any plans, projects or processes start completely afresh. There has often been some previous plan or project and, for many participants, previous engagement. With luck, there may be little or no bad experience from that previous initiative but, more commonly, especially where one is trying to raise standards from previously poor involvement, there is a need to prepare key people for a new and different process. In Chapter 3 this was referred to in terms of the negative 'baggage' that people bring with them from past experience. Such concerns need to be drawn out, understood and dealt with or they may linger and damage the forthcoming process.

For relatively low key issues, a single session that brings together the key people can be all that is necessary. Such a session can draw out the concerns, clear the air, deal with misunderstandings and begin to set up some good practice requirements (principles) that would please everyone and lead to their commitment to the next process.

If concerns are more deep-seated, or levels of mistrust are so high that key people are not even willing to share a room with each other, the usual approach, familiar from international mediation, has been mentioned already: 'shuttle diplomacy'. This involves someone from well beyond the issue seeing each key group in turn, hearing their concerns and, as above, hopefully clearing the air and beginning to set up some good practice requirements for the next shared process. In particularly tense situations the order in which this is done can be important, so give real thought to that. Things may also not be resolved in just one round; it can be necessary to run two rounds (or more) so that those who have been seen first, second or third have an opportunity to hear about what others have been saying (although what is or is not shared also needs great care).

All this is valuable but something else can be done that potentially seals in any positive outcomes of improved trust or willingness to try again. That is to use one of the principles introduced in Chapter 2 and involve the key people in designing, and perhaps managing, the forthcoming process. It is almost classic that if they report that they were not listened to, there was no feedback and groups A, B and C were never invited last time, then giving them a chance to design a process that deliberately remedies all of those things the next time is remarkably effective. How to do this will be picked up shortly.

The building blocks of a process

Any collaborative process needs a start (nowhere near as easy as it may seem) as well as an end (only a little less difficult). It also needs stages along the way at which to take stock, report back, make changes, perhaps even celebrate. Processes also never exist in a vacuum. There will always be some parameters of overall time and timing (completion before a certain date), resources (in all senses, not just monetary), other influences (a forthcoming election), ensuring momentum (not too fast, not too slow), lead-in times (timely warning of event dates to participants) and so forth, every one of which needs to be taken into account.

There are also points to watch for about timetabling a process because it is not all about the key events or activities:

- The majority of the time in developing any plan or project will almost always be taken up by the technical team in advancing their work (testing options, checking traffic flows, assessing market capacity and so forth).
- Reports need to be prepared and circulated, giving people time to consider and probably share them with others. A rule of thumb is that representatives from community groups, NGOs and even some authorities, require a minimum of one month following a workshop to report back to their group and prepare ideas and input for the next workshop.
- Further information will probably need to be sought, and results from each stage have to be analysed and prepared for the next stage. It is bad practice to fix the next meeting without knowing when key information for that meeting is available.
- Specific people who did not get involved may need to be contacted and reassurance is needed for those who did and are perhaps now cautious about their involvement.

- Some process changes may need to be considered and agreed.

The various participants may all agree that they can, for example, meet once a month but all the factors listed here may well make that impossible and such constraints should be given serious consideration when considering time and stages.

To make sense of this, different process designers use different formats, grids and diagrams to design, describe and communicate their processes. None is ever completely right, not least because no single diagram or set of headings can possibly capture all the issues and ideas at play in a complex process. Based on this author's experience, the most important features or building blocks around which to develop some sort of flow diagram are the following.

The first three building blocks described here might be thought of as external to the process itself but they can affect, sometimes even determine, the main process.

External constraints

Some planning (and other) situations are tied completely to statutory processes, for example, preparing a formal plan. This can constrain in terms of what can and cannot be addressed, the exact standard of evidence required, the stages through which to pass, the timeframe for completion and even the detailed nature of the finished report/plan. Some processes can be statutory but be far more open, for example, a regeneration framework (as for Waterton). Some may be almost completely non-statutory but constrained by other factors, for example, when a developer needs to secure permission by a certain date to accord with an option agreement, or when there is a requirement to submit highly prescribed reports.

Communication

At one extreme end of collaborative working, for example some international mediation, a process only works if it is completely private; we (the public) only know it has been going on when it is resolved (or not). Situations such as those central to planning and environmental change are probably at the other end of the spectrum in that they are inherently public and need to be played out as openly as possible with all consultees having an opportunity to be involved. There is therefore a need for carefully managed communication beyond the main core work and core group of any project, and especially if invitation-only events are part of a process.

This might be thought of as a public relations task but, as explained in Chapter 3, great care is needed with this because classic PR is partial and committed to one client's approach and solutions. As meant here, communication is about helping to raise awareness of a project, share information, encourage participation, feed back stage outcomes, receive information or ideas (mainly proactive), and occasionally deal with gossip or misinformation (reactive). It is not and must not be about advocating a particular solution or advancing just one agenda.

Awareness-raising

Planning and environmental situations may be inherently public but it does not automatically follow that people will know about or understand that change is on its way and may affect them or appreciate that they can get involved in the process. That has already been explained in terms of immediacy – 'what has a waste strategy got to do with me?' – but it can be even more basic if people do not even know that a strategy, plan or project is being prepared, even one likely to have an impact outside the front of their house.

In addition, even when people know about some change and understand its relevance to them, they may still not get involved because 'if they have never taken any notice of me before, why should I bother now?' And if an invitation to become engaged follows several others on other local issues then cynicism and fatigue can easily, and understandably, set in.

Breaking this vicious circle on single projects can be extremely difficult, suggesting a need for a variety of forms of awareness-raising. This might be having the project mentioned in the media but more is usually needed. Working with a key group of stakeholders is particularly valuable because, as trusted people locally, they can take key messages to potentially very large numbers (and be listened to differently). Working through schools can be valuable, as can 'piggy-backing' existing events such as a community festival. Most importantly, awareness-raising, though mainly an up-front activity, needs to go on throughout a process to slowly build knowledge, increase confidence about being listened to and engage even more people.

The next building blocks described here deal with the main approaches to take or types of event/activity to use. The categories do not divide up anywhere near as easily as the terms suggest but this division has shown itself to be a pragmatic way of avoiding a morass of potential choices.

Collaborative design and management

The third principle of effective engagement in Chapter 2 was 'Agreed process', i.e. encouraging key players to join in designing a process or at least having an opportunity to comment on and endorse it. This is usually done by bringing those key players together in a workshop. It is difficult to over-estimate the value of this but two cautions need to be raised:

- It is not easy at the outset to identify key players and it can be risky to try to do so because others may raise concerns later in the process.
- Few people have any experience at all of intense engagement or collaborative working so it can be challenging for them to contribute in an informed way.

Both issues suggest that care must be taken not to over-value the outcomes of an early, shared design session. This is best done by being attentive to new issues, people and ideas as they arise and being willing to adjust any initially agreed process. The box below describes what took place for Waterton.

> For this project a workshop was held at the very outset with around 20 people. Participants included planning officers, development agency representatives, elected members, a Chamber of Commerce representative, someone from the local college, someone from a major local business, officers from the County Council and some from local NGOs. The session was very interactive and covered the following:
>
> - **Introduction and a 'preface'.** The 'preface' was an invitation to people, as they arrived, to place a tick, cross or question mark on a big sheet listing engagement principles to show whether they agreed or not with each principle.
> - **Principles of good engagement.** This was feedback, discussion, elaboration and agreement of the principles.
> - **Key issues.** Although the session was about process, this active session on issues helped people to focus in on more familiar things while at the same time giving them hints about the next stages of the session.
> - **Scope of the engagement.** Some parameters were suggested – the framework is to cover X area and must deliver Y new houses etc. – and then discussed, elaborated and agreed.
> - **Consultees/stakeholders.** This was an active session over a break when people listed all the groups, organisations, agencies and people that they thought ought to be involved.
> - **An outline overall process.** This was presented as a blank framework* with some initial post-its added to suggest a workshop here, a meeting there, an open event there etc. In discussion, moveable notes were then added, removed, changed and moved around on the framework.
> - **Agreeing a Steering Group.** Using some agreed criteria, for example that the group should be no more than ten people and only have two councillors, the group agreed a list and some broad terms of reference for the group.
>
> * The framework is introduced later in this chapter, as is the agreed Waterton diagram/process.

As in the Waterton example, another way to avoid later problems, often valuable in its own right, is to establish some form of 'Consultation Steering Group'. This provides continuity from any opening workshop and engages the group members not just in advising throughout the project (for example, on accessing groups not responding) but also in ongoing discussions about adapting the originally agreed process.

There is no simple rule for the size of a group, although 8–12 has worked well for this author. The group should be widely representative, typically officers, members, business interests, voluntary/community interests and, if a specific site is being planned for, some neighbours to that site. Any formal or informal terms of reference should make clear that the group is not there to comment on the content (plan or design) but *solely* on the delivery of the engagement process and on the validity of the final results in relation to the engagement.

This last point is important because, at the end of a process, the Steering Group can be asked to endorse the final report of engagement (see Chapter 7). If this can be achieved, it adds significantly to the status and value of the report because it is no longer just something done by the process manager or commissioner; it has the community's sign-off.

Activity 'in-depth'

To be successful, collaborative, deliberative or dialogue approaches depend almost entirely on face-to-face events where people engage together. They are also only fully effective if done as a series; even the best single event has limited value. This is because those involved are usually representatives (stakeholders) who need to report back, because initial activities generate a need for more information and because initial options need to be tested. It is also because issues, ideas, options and balances need to be reflected upon by participants in a way not possible during an intense collaborative event. This has already been described as work in-depth and some engagement projects have done nothing but this; for example, the final Thanet Coast Management Plan emerged from four day-long workshops spread over ten months with a group of around 70 stakeholders.[2]

Some work in-depth is the indispensible, core element in any good engagement process if truly wide and informed support is to be secured. However, that does not have to mean four day-long events with 70 or so people, especially if the considerable resources involved (for all) take away opportunities for others to get involved. But it does mean planning carefully so that the results of in-depth work link into (and/or back from) 'in-breadth' work as below. (Some in-depth work can also be done 'at distance' as explained later in this chapter.)

Activity 'in-breadth'

This is about providing opportunities for anybody and everybody to make a contribution. However, work in-breadth is not, certainly in earlier stages, about preparing static exhibitions or displays that people simply look at. There are almost endless ways of running open events or activities that are genuinely interactive and they are what is primarily meant here by in-breadth work. Displays of emerging or almost final plans or proposals have their place when most engagement work is finished but, even then, there should be opportunities for people to comment.

It is of course naïve to imagine that everybody would come to any event for anything other than the smallest local project, perhaps not even then. So the choice is not always to offer something to which people *come*, because that inherently limits take-up. The other obvious choices are to *go to where people are* (using a stall in the high street or mini-questionnaires in local shops), '*piggy-backing*' a local event, or *meeting halfway* by using the many forms of social media now available. In addition, the activity need not even be a single event. Roadshows can be used to go round to people, and other creative (some might say sneaky) methods can be used where people may not even be aware that they are being consulted.[3]

Be cautious, however, because most in-breadth methods are predominantly individualised; they do not allow for any, or at most very little, interaction between participants as is central to core collaborative approaches. (One can, however, run a short in-depth workshop for a dozen or so people during an open drop-in.)

Balancing in-breadth and in-depth work is also important because both are needed, even if depth work forms the core of any process. To avoid being swayed by those who think that all that matters is the numbers who contribute, take a simple comparison:

- Two hundred people come along to a drop-in over two days, spending an average of 30 minutes at the event (breadth); a total of 6,000 individualised consultee minutes.
- Twenty-five people come along to three workshops of 150 minutes each (depth); a total of an enormous 11,250 interactive and deliberative (*not individualised*) minutes.
- What's more, those 25 people at the workshops can share issues back with many more in the groups they represent. In one case 24 group representatives accessed almost 2,000 people.

Outreach

Inviting stakeholders to workshops usually succeeds, though some groups may want to send all their members or need persuasion to send just one or two. Good awareness-raising, well located and well promoted events and sessions that go out to people can all be valuable ways to get through to others. However, there are sections of the community who are, as described in the previous chapter, hard to reach. Even the best awareness-raising, well located event etc. will not get to these groups or attract them to join in, so some sort of special, targeted effort is needed.

This area of work is often called 'outreach' but this one book cannot offer real detail on how to do it successfully; in fact, nobody yet seems to have found a miracle set of approaches or methods. Two basic and useful points can, however, be made. First, rather than trying to do this oneself, it is nearly always better to try to get to people in the various hard to reach sectors through others who are specialists or who have direct contact: A youth worker, a social worker, a teacher or doctor, even via people's places of work.[4] Second, the real aim is not just to contact them and seek their contribution in isolation. The real aim is to get them to join in at the heart of the process, to become a stakeholder at workshops or join the in-breadth activities with others. That can be really difficult but it should at least be kept in mind as an aim.

Stakeholder analysis

Applying the guidance in Chapter 3 on stakeholder identification can generate an enormous list of possible stakeholders or consultees (the one given for Waterton was only a selection). This is the point at which one often hears cries of despair about how to manage a long list sensibly. This is where **stakeholder analysis** comes in, where it relates back to ideas covered earlier in this book about different levels of engagement and where it relates to choices such as in-depth or in-breadth work.

To identify different categories or levels of involvement for stakeholders, the first stage is to agree the most important criteria for involving them. For example:

- Is the aim an equal balance of coverage from all categories or sectors?
- Are there some categories, even some groups or people, who are clearly top or low priority?
- Are there some who can be included or engaged best through work in-breadth and others who must be at in-depth events?

A brief look back at what one is trying to achieve with the process, at the preparatory work (information assembly etc.) and at the emerging list of stakeholders will help with criteria selection. There is then an extremely practical method for organising the remaining list; a method that can also be done very easily and usefully with any Steering Group. Assuming a workshop setting:

- Start by taking a sheet of flip chart paper and on it draw a grid, as on the diagram in Figure 10.
- Now write the names (or get participants to write the names) of each separate potential stakeholder on a separate Post-It note.
- Try to agree where to place the Post-It notes on the grid (or let participants do it).

Post-It/moveable notes are placed vertically according to how much *influence* those named on the note are judged to have over the process and its outcomes and horizontally according to the degree those named will be *affected* by the outcomes of the plan or project.

This is not, nor should it be, an instant task. One stakeholder may only have influence at one particular stage, another may have little influence formally but a great ability to lobby and be listened to. It is also important not to rush this process because different people in a group may have different perceptions and experience; the discussion that then takes place is important for all.

For an example of how the result might look using a few of those from the Waterton list, see Figure 11.

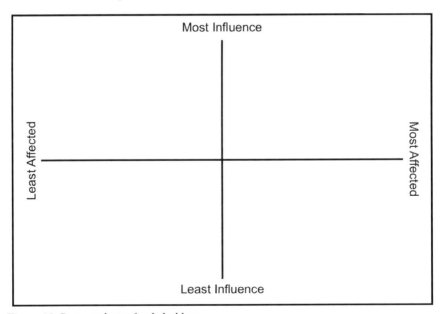

Figure 10 Basic analysis of stakeholders

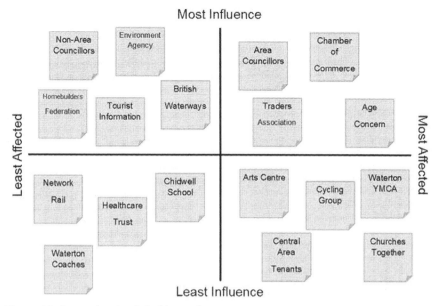

Figure 11 A completed stakeholder analysis

Now one can use the result to provide some key pointers to who to engage, when and how. Most importantly, a link can be made to the typology of approaches used in Chapter 2: **informing** to **consulting** to **involving** to **dialogue** (engaging). Applying this to the diagram suggests the basic (just the basic) approach to those in each quadrant as follows (and explained in Figure 12).

- Some stakeholders (bottom left quadrant) have little influence and would be minimally affected. They can simply be kept informed regularly so that they can contribute to in-breadth activities if they wish.
- Some (top left quadrant) are quite influential but only affected indirectly or minimally. They can be linked in to a planned programme of contact so that they can be offered input at what they consider to be the right time for them and on appropriate issues. They need to be contacted to agree this.
- Those in the bottom right quadrant may be very affected, and directly, by the plan or project but have little influence over final decisions. They need to be involved in a variety of ways, which may include invitation to in-depth, dialogue events but certainly invitation to all in-breadth activities.
- The key group in terms of collaborative planning and rigorous dialogue is those placed in the top right quadrant. They need to be firmly, thoroughly and regularly engaged, probably in approaches that are about face-to-face dialogue and deliberation. They should, however, also be invited to in-breadth events because that can help to engage all those they represent.

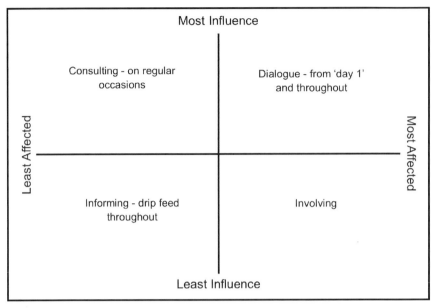

Figure 12 Approaches derived from analysis

As well as being a practical and potentially collaborative technique, this can remove many of the fears about how to address long, challenging lists. It can break down a long list very quickly into more manageable numbers, also based on a rationale for how best to deal with those in each quadrant. It should go without saying that the technique is rough and ready; it does not give final answers but it moves the thinking several stages forward. There is also a point to make, given this book's emphasis on collaborative working, that there can be good reason to widen out the list of those in the top right dialogue quadrant, certainly to consider the inclusion of some from the top left and bottom right quadrants.

From long experience, the Influence/Affected axes are usually the most informative. Other criteria and hence axes can also be used if particularly relevant, for example:

- **Representation**: How many people, or even other groups and sectors, a stakeholder might represent.
- **Interest**: Whether the interest is in just one aspect of a project (for example traffic) or all.
- **Geography**: Physical proximity to an area or site.
- **Role**: Formal/statutory (see Appendix 2) or informal.
- **Ease of access**: Some will be easy to contact, others will be part of the 'hard to reach' list.

From building blocks to a process

The build-up so far in this book has been deliberately slow because of the real importance of careful preparation, but it is now the time to take a leap and look at an overall process design. Given the challenges of explaining design, the best way to do this is to look at one specific whole picture ('menu') then draw the general points from that. The Waterton Regeneration Framework is the example used here. At the process design workshop described earlier in this chapter (see box on p. 72) the group was presented with a very large version of the blank grid as in Figure 13.

No apology is made for the fact that this diagram does not replicate the exact terminology as introduced by this author. It is important to use diagrams that are appropriate for use with people and groups in many different settings, so this one is deliberately slightly different to reinforce the point about application to any local context. This basic model has proved its value in many projects by being practically useable by project leaders but also by ordinary people; most professional project management grids are totally incomprehensible to ordinary people (and also to some professionals).

Figure 14 is a very simplified version of the diagram for the Waterton project. This was developed at the process design workshop and resulted from lots of discussion, added Post-It notes, removed notes, amended notes, moved around notes etc. And it changed again, quite rightly, once work proceeded.

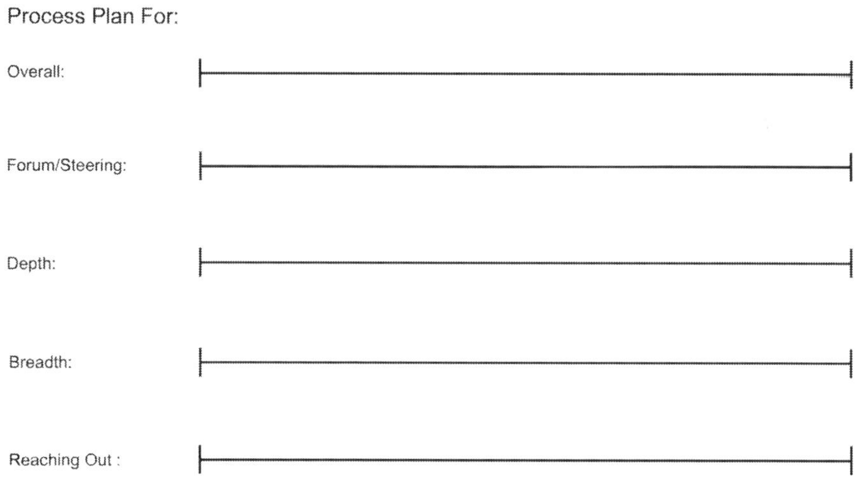

Process Plan For:

Overall: |——————————————————————|

Forum/Steering: |——————————————————————|

Depth: |——————————————————————|

Breadth: |——————————————————————|

Reaching Out: |——————————————————————|

Figure 13 A blank process plan

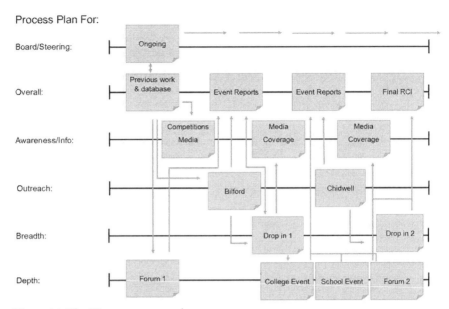

Figure 14 The Waterton process plan

The main lines and boxes in Figure 14 can be explained as follows:

- **Board/Steering**: This was ongoing because the diagram was produced early in the project before any dates had been set for Steering Group meetings. It was also recognised that those dates should not be determined solely by external pressures (other than a broadly agreed end date) but by stages in both the technical flow line *and* the engagement flow line. As work proceeded, dates were added, in particular for the group to discuss the event reports (as on the line below, which corresponded with technical stages). This ensured a managed opportunity to review and make any necessary changes.
- **Overall**: The first grey box in this diagram (Previous work and database) suggests another reason why some lines could not be elaborated at the outset. As stated already, projects rarely start afresh; they have histories and there may have been previous consultation. The same is true of the database of consultees; as outlined earlier, this can be small and out of date or it can be full and topical. When the diagram was done it was not clear whether what had happened before would be a benefit or a burden. Note also the inclusion at the end of the line of the final Report of Community Involvement (see Chapter 7).
- **Inform/Awareness**: This line combines the Awareness-raising and Communication building blocks. Awareness-raising was regarded as extremely important in this project because there had been a very poor

history of community engagement in Waterton. Communication throughout, especially to promote the in-breadth events and to feed back the results from them and the in-depth events, was also important. How this was done is explained in the next chapter.
- **Outreach**: Once again, given the poor take-up of engagement opportunities in this town, outreach work was also important, especially, as shown, as a precursor to other main events. At the start, the main focus was on specific neighbourhoods and communities where community development had been, and still was, taking place. However, young people were soon identified as a key 'hard to engage' group so other activities were added later to this line.
- **Consult**: This was the group's preferred term for the in-breadth work; they felt that it would mean more to people. It is explained very easily, as suggested by the two boxes. Two open, public drop-in events were run. One was in a very central location which gave access to a very large number of people, including many who would never otherwise have made a point of coming to such an event. The other took place in a tent at the town's main annual festival, again accessing many people who might otherwise not have come and even some from outside the town.
- **Involve**: This was the group's preferred term for the in-depth work and is again explained simply. Two workshops (Forums) were held for invited stakeholders. These took place at key design stages and before the drop-in events so that the material used at the drop-ins was as much from local people as from the technical team. Two other events were shown on this line even at the early stages before young people were highlighted as a key group, namely workshops with young people from the local college and from local secondary schools.

The very bottom box must also be mentioned. The regeneration project was under way at the same time as several other projects. The regeneration project team tried hard to find out about all of these, to contact their project teams and to coordinate any consultation. This proved very difficult and yet, from a public perspective, a reasonable assumption is made that key agencies work together. As a result, people were surprised, and sometimes annoyed, when they were consulted on other projects having just been consulted on the regeneration work; an example of understandable 'consultation fatigue'. All the regeneration team could then do was to manage this and limit its impacts.

The arrows between boxes are also important because they highlight key sequences, for example, Forum 1 first, some Outreach second, Drop-in 1 third and College event fourth. Results or issues from one were used to kick-start or inform another.

Delivering through stages

More points could be made about choices and options around producing and agreeing such diagrams than can possibly be covered here but, from experience, three are especially important. They relate to the main stages in any process (roughly): Starting, advancing and finishing. Each needs real attention because they do not just happen because some clever methods are used. The idea behind this comes from the work of Sam Kaner (Kaner, 2014), see Figure 15.

Starting (or starting again)

According to Kaner it is very important in any process to get people firmly and clearly *into* the process, ideally dropping the baggage or prejudices left over from previous experiences and understanding (as much as possible at this early stage) what they are about to get involved with.

Starting can be difficult because, when working with a process planning group, the longest discussion is usually about the very first step. At one level it seems best, one might say democratic, to start with a completely open launch event (in-breadth) for everybody, and this is what a group will normally suggest. But remember, if most of the people you aim to attract to such an event are not yet aware of a plan or project and its potential implications for them they will probably not attend, however well promoted your event is, and time and effort could be wasted.

Though it may seem counterintuitive, it can be better to start with an invitation-only stakeholder (in-depth) event because, if successful, those involved are probably going to be the greatest potential allies in terms of going back out and raising local awareness ready for the first open (in-breadth) event.

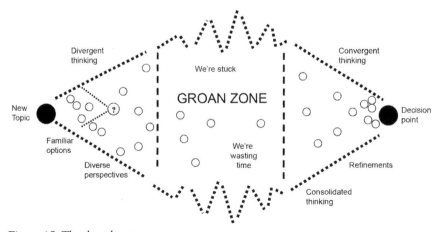

Figure 15 The three key stages

Furthermore, material produced from this first event, if used at a follow-up public event, is less likely to be seen as something from 'them', but as something that comes from 'people like us', because it already embraces local issues, concerns and ideas. That sense of independence can be remarkably effective in helping to engage others and build confidence.

Within this starting stage there is also a need to ensure, as far as one can, that everyone is appropriately up to speed on planning and development procedures, even on things as apparently straightforward as being able to read plans. On some occasions an enjoyable and informal initial 'induction' event has proved useful. (Such things are best not called 'training', though they are!)

It is also crucial, whatever activities are done in the initial stages, to explain to people the overall, properly planned process and that opportunities will emerge to deal with particular issues at the most appropriate time. This can help them to make decisions about their (or their colleagues') future involvement. A broad point of principle might best be addressed at a first event, details of design at a far later one. This is also important because too often when only one event is offered or a process is not made clear people naturally tend to want to ask questions and see answers to every single issue at the very first event as they imagine it is their only opportunity to do so. This becomes far less of an issue once they see an overall picture.

Finally, for the initial stages, particular effort needs to be put not just into the technical content but also into establishing relationships, building trust and enabling people to work well together. Not that such factors are unimportant at all other stages but, if not done early, this can limit or slow progress at later stages.

The groan zone

There is often a quite long middle period when there can be questions about progress, frustration, a feeling of going nowhere, even a need to go backwards, for which Kaner uses the highly appropriate term 'the groan zone'.

It is a fundamental principle of consensus building that solutions should be left as open as possible as long as possible. This means that there is not any easy, sequential, A to Z process from issues to vision to options to solution. It is all messy and opaque and is often known to produce 'groans' as people struggle to balance their varying opinions in relation to interim solutions that stubbornly seem to never offer easy choices. It is also almost always the case that initial work changes some technical or content parameters and even the agreed process. In the Waterton case *three* process diagram variations were produced with Steering Group help because what was actually happening and what was appropriate was changing. This can also involve going back through key stages or issues once or even twice; the process can quite rightly be iterative, not just sequential.

Not surprisingly, this can be the most difficult stage, with no apparent end, successful or otherwise, in sight. This is another reason why providing the clear, strong landmark of an (ideally agreed) overall process is so crucial, even if it then changes. There are three basic points to make about finding one's way, and enabling all others to stay committed and find their own way, through this awkward stage:

- First, what *not* to do. If at all possible, and this can be difficult if certain timescales are rigid, avoid rushing, foreshortening or curtailing any key stage or activity.
- Second, do everything that you possibly can to make it crystal clear at the very outset that delays may happen, things may have to be revisited, it won't necessarily be easy and quick; i.e. avoid surprises and a sense that something has gone wrong. Any Steering Group can be particularly valuable for helping to get this message across because they should be aware of the issues. Interestingly, from long experience, few people are ever surprised by this; as someone once said about back-tracking, "Nothing new there; things are always like that, aren't they?"
- Third, use a simple but extremely effective technique called 'tracking'. It is quite understandable that people get engrossed in the content of some plan or project and forget where they started, where they have got to and where they are going. Tracking (see Chapter 6) simply means telling or reminding them, very regularly, how far they have got along the agreed process, what stage they have reached and what still needs to be done, by when etc.

Finishing (or closure)

This might seem easy after the groan zone; a nice gentle flow towards developing a plan or project in detail. However, it is all too often only when a specific solution begins to come into sharp focus that some stakeholders start to worry. This can be an understandable result of a lack of appreciation because it is only when detail starts to emerge that non-professionals may be able to see the full implications of something that they had supported in principle up to that point. Unfortunately, even if high levels of trust have been built up, raising objections can also be deliberate and hence damaging. That is because some people wait to see what is happening and then object once an emerging solution that they do not personally support begins to gain wide support from others.

If anything, one needs to avoid what can seem like an easy rush to conclude. This is a classic point at which to look back over the whole process, revisiting and reconfirming initial objectives (for both content and process), the criteria or priorities set in early stages, the options considered but rejected and generally undertaking a thorough test of what now appears to be the solution. This can produce more groans but it always helps to generate some final, detailed changes and improvements to what is emerging.

This is also the stage at which the enormous medley of available methods starts to run out. As will become clear in the following chapter, there are many methods available for the starting stages but far less for the groan zone or finishing stages.

Consensus without collaboration

This odd title refers to a form of collaboration that can lead to consensus solutions even if it is not face-to-face, as promoted in this book. It refers to collaboration that happens 'at distance', i.e. those involved may never meet together. This approach was first developed in its detail (though others now use it) by an organisation called Dialogue by Design.[5]

Consensus without collaboration essentially involves electronic communication with key stakeholders (it is not really appropriate for wider public settings), often on issues that impact over a considerable geography, even internationally, thus helping to avoid too much long-distance travel to get everybody together. In one case the approach was used with representatives from countries all across the planet.[6] In another, for a housing project in Salcombe, it was important to find ways in winter to contact large numbers of summer holiday home owners not permanently resident and only available by email or letter.

This might seem familiar, after all several websites already offer interactive opportunities. But beware, because almost all of these sites are based on one-off exchanges using simple questionnaires and they are almost always individualised. Online collaboration or consensus-building, though seemingly based on questionnaires, nevertheless operates in a very evolving, cumulative and transparent manner that has some parallel with face-to-face dialogue managed through a series of workshops. Typical process stages might be:

- An initial contact with a first group of stakeholders to develop an agreed and longer, wider list of participants.
- A first main stage to scope issues on a particular topic and feed back some sense of agreement and priorities.
- A subsequent stage to ask people to rank the issues.
- Stages to seek possible solutions or parts of solutions to each or all issues, which are again shared and ranked.

- Development of options or a preferred solution, which are shared and evaluated against the initially agreed issues or criteria.
- Areas of disagreement explored with some or all and hopefully resolved.
- Further stages that proceed towards final, shared agreement.

Once again, this offers not just one 'ingredient' (a questionnaire), but a proper 'menu'; a planned and managed, consciously designed overall process.

Key messages

- Developing a coherent, integrated, cumulative *overall process design* is absolutely key to successful outcomes.
- Within any process, varying approaches should be used but dialogue, i.e. face-to-face work, should be the core or the pivot wherever possible.
- Processes are about getting the right people together in the right place at the right time, using the right methods that add together to cumulative good effect.
- Once the stakeholders have been listed, make the list manageable by working out who may need to be engaged, in what way, at what time and on what aspects.
- Ideally, design the process interactively with others.
- Produce some form of summary process diagram which is of value to the process managers in running the process but also understandable by non-process people.
- Be aware of the different approaches or styles needed at different stages – start, middle (groan zone) and end.
- Be aware of reasons why a process may need to change and change it appropriately.

Notes

1. For example, Enquiry by Design as promoted by the Homes and Communities Agency and others or Future Search (Weisbord and Janoff, 1995).
2. Twenty years later topical issues for the Thanet coast are still being addressed through collaborative working.
3. See Wates, 2008 and 2014 for a superb medley of good methods.
4. In one case this involved workplace lunchtime discussions with the 'bribe' of free sandwiches.
5. Dialogue by Design: www.dialoguebydesign.net
6. This was for the UK's many small dependencies scattered across the globe.

References

Driskell, D. (2002) *Creating Better Cities with Children and Youth*. London: Earthscan.
Environment Agency (undated) *Building Trust with Communities*. Available at: http://repository.tudelft.nl/assets/uuid:6c88a7b9-1478-435d-ad58-95b7b59868b2/ComCoastWP4-07.pdf. See also: www.environment-agency.gov.uk
Kaner, S. (2014) *The Facilitator's Guide to Participatory Decision Making*. San Francisco: Jossey-Bass Business & Management Series.
Wates, N. (2008) *The Community Planning Event Manual*. London: Earthscan.
Wates, N. (2014) *The Community Planning Handbook*. London: Routledge.
Weisbord, M. and Janoff, S. (1995) *Future Search*. San Francisco: Berrett-Koehler.

5
Delivering into Detail

Introduction

The three sections in this chapter move from overall processes or 'menus' (as in Chapter 4) through planning specific events to planning specific sections of, or sessions in, an event – the 'recipes'. To design any session one then needs to know what methods can be used – the 'ingredients'.

The key caution about sequence must be mentioned again because, until one has good experience and knowledge of various methods and techniques, one cannot decide on or start to plan any specific event; it is all circular. Be aware also that there are many ways to manage an event (and many different sorts of event), many possible ways to manage a session and almost endless methods and variations to them. Those described here are the more generic, typical and proven ones only and, although there is guidance here on activities and formats for interactive drop-in sessions etc., the focus is mainly on collaborative, dialogue activities which bring people together.

This chapter works its way from processes to methods through three stages, each with its own section:

- From process to events
- From events to sessions
- From sessions to methods.

From process to events

Having warned about endless choices, this section focuses on just two that might be regarded as generic. The first is an invited stakeholder workshop, which is a fully collaborative event. The other is an open 'drop-in', which is an interactive if not fully collaborative event. Several points apply to both:

- Be really clear about the basic purpose of any event.
- Identify the key stakeholders or invitees (that includes 'everybody' for open events).

- Be clear where the event sits in the overall process and any implications arising from this.
- Check an appropriate venue.
- Check that the event timing does not clash with anything significant: school holidays, a council/board meeting or some other consultation event.
- Make a (very) provisional decision about overall times and timing.

Do not make final decisions on any of these until after completing *all* the following stages.

Workshop

In terms of the scope and format of any collaborative workshop there are seven main steps as follows:

Step 1 Establishing purpose, outcomes and related outputs: From work done to develop the overall process, there will be a defined purpose and one or more desired and feasible outcomes and outputs. Note these and keep coming back to them because session design will evolve, perhaps providing suggestions or amendments to the outcomes/outputs, and they will be needed to form part of the event record and for later evaluation. Some *process* **outcomes** might be:

- Everyone has understood and agreed to the purpose of the event.
- Everyone has understood the role of the facilitator.[1]
- Everyone has had an opportunity to hear ideas, issues etc. from others and add their own.

Some *content* **outcomes** might be:

- There is now a wider and fuller list of issues to address.
- Agreement on the key issues.
- Some initial possible solutions to those issues.

The key **output** should be a full event report, certainly circulated to all present and probably shared more widely, for example, via a website. Other outputs could be a draft overall policy (for a plan) or sketch layout (for a development).

Step 2 Checking timetabling and timing: When to hold different sorts of event is covered more fully in Chapter 6 but the key choices boil down to weekday, weekend or evening. How long should the workshop be?

- Full day or half-day if during the day, two/three hours during an evening. (Two- or three-day sessions cannot be ruled out but have so many practical limitations that they are likely to be rare.)
- Fix this in consideration of what amount of time the listed stakeholders might be willing to give and when.
- The above will suggest start and finish times but the other constraint for stakeholders is any necessary travel time (and, for example, the availability of buses home at night). For others, such as older people for example, the time of year and time of day is important in terms of feelings of comfort and safety. Once again, their criteria need to hold sway.

With provisional times in mind, check them with venue managers and factor in to any booking at least one hour at the start to set up (ideally two) and perhaps 30 minutes at the end to clear up and run a quick review.

Step 3 Blocking out timetable fixes: Consider the following when planning the timetable:

- Might it, for example, be a clean start at 4pm or a 'join us from 4pm for some refreshments and a prompt start at 4.30pm'? (The latter always ensures more people are present for any formal start.)
- If the intention is for people to do an 'arrivals task' before actually starting (this will be explained later), allow for that in the timings and what is told to participants.
- A break of some sort every 90–120 minutes is good. With a small group, a short comfort break of ten minutes might be enough but a larger group would take perhaps 15/20 minutes. Add at least five minutes, maybe ten, if the break also includes drinks etc. Though working breaks to avoid stopping a session are possible for well-motivated groups, they still add time.
- If there are to be refreshments on arrival, allow time for that. If refreshments are to follow an event, allow for this in overall venue booking time. Always avoid formal seating and served courses if food is provided; that always takes too long. For day events, 45 minutes for a lunchtime break is a minimum. Check all of this with venue managers.
- There must be some form of opening session, it can be as short as five minutes, to introduce people, locate the toilets and fire exits, set the scene and describe the process. At initial events in a complex process this can take more time, maybe 15 minutes or more.
- Always build in time for a closing session, even if only a few minutes, to bring the event back to ground, ensure people are clear about progress made and set the scene towards any next stage/event.

People rarely arrive on time (even with the temptation of free food), they can take longer to move around and, on almost every occasion, any planned programme changes. So leave some slack and consider also having a *private* programme as well as one for participants. (For example, start time for the private programme might be later than participants are told.)

Step 4 Producing a basic template: Drawing from the steps outlined thus far, one can then lay out the basic private programme for an event for a full day workshop as follows:

9.00	Convene, drinks and arrivals
9.35	Start and Introductions
9.45	*Session 1*
11.00	Break
11.20	*Session 2*
12.45	Lunch
1.30	*Session 3*
3.00	Break
3.20	*Session 4*
4.00	Closing
4.15	Finish

Note: Do not 'ink this in' yet.

Step 5 Identifying key process and content points: Whoever plans an event should have control over the process but the content, what it is and how it is best presented for people to use it in their collaborative work, also needs to be planned with extreme care because, to be effective, it infuses everything that happens. However, it can sometimes be difficult to influence this because it is seen as the territory of the technical team.

This is particularly important and relevant if there is to be any form of presentation by a member of the technical project team. What needs to be covered in a presentation and how long it is to be may be clear, even agreed, but very few speakers keep to time and, what is worse, some even fail to keep to agreed content! Any private programme needs to allow for overrun and the introduction by the speaker of some different, irrelevant, even worrying points. Strong reminders to presenters on the day and countdowns of one or two minutes during any presentation can also help.

Step 6 Achieving the outcomes: Knowing the time available for interactive work should now make it possible to draft out initial ideas for what might be done in what sort of way in each of the available sessions (steps 1 to 4). Once the detail of this is in place (see following sections) a programme can be drafted for participants. By now the private programme will be in real detail but it can be counter-productive to share that with any participants. As one never keeps exactly to time it is important to avoid a situation where people feels that something has gone wrong if a session ends five minutes later than on their programme. With that in mind the draft programme outlined in Step 4 might look as follows to participants:

9.00	Please join us any time from 9.00 for some refreshments and to complete a small task
9.30	We start promptly at 9.30 with some formal Introductions
	In our first working session we will …
	There will be a refreshment break before the next session
	In this second session we will …
12.45	Lunch will be served around this time
1.30	In this third session we will …
	There will be a refreshment break before the next session
	Our final session will …
	There will be a short closing session, looking ahead to the next event
4.15	Finish of workshop.

Despite the focus on flexibility, overruns and adaptation, it is extremely important to try to finish on time. People value this, it creates a sense that those managing the event know what they are doing and there can be practical issues to consider too, for example, about transport home.

Step 7 Finalising the programme: With the framework in mind, planning now shifts to session design and method selection as in the following sections. Once those are decided, go back to check whether each specific session can be managed in the time allocated in the draft programme. The example programme here shows equal time for all sessions but only rarely will this be the case. That requires returning to the draft programme and adjusting it before it can be considered final.

The Waterton project provides a useful example of a finally agreed and filled-out *private* workshop programme, as follows:

> This was for the opening workshop to be attended by a very mixed group of around 20 key stakeholders.
>
> | | |
> |---|---|
> | 5.30 | Arrive to set up – project team to arrive for 6.15pm |
> | 6.45 | Tea/coffee and biscuits as people arrive. |
> | 7.05 | **Introductions** |
> | 7.15 | **Issues and Aims.** There will be a series of wall-sheets covering a range of issues* and project aims. Participants will walk around ticking, querying and adding. This gets them up to speed and enriches the lists. Quick plenary feedback and review. |
> | 8.00 | **From Early Opportunities to Possible Approaches.** We will make it clear that: (a) this is not a time to be *choosing* options or 'the' option; and (b) they are not determining anything, merely sharing their views. The team will set out very initial suggestions for the key opportunities and some general examples of the sort of approach taken to option development. Although these suggest a main focus, the aim is to ensure that, in their group work, people address as many as possible of the issues and aims now on the wall-sheets. Mixed groups, each with access to a team member, work on their allocated approach through annotated plans and flip chart notes – housing site here, retail there, gateway, job creation opportunity etc. |
> | 9.00 | **"Going through to the first round are ..."** Walk around to look at what has emerged, then hold a plenary discussion of which approaches or variations thereof (or new ones) might be taken forward to be developed up as genuine options for technical testing and drop-in consultation. |
> | 9.25 | **Where Next and Close** |
> | 9.30 | Clear up, collect results, quick team debrief. |
>
> * 'Issues sheets' are explained more fully in the next box and later in this chapter.

Open drop-in

As with workshops, explaining the myriad ways in which to run a drop-in, display or exhibition is almost impossible. So this section starts with an example (following box) based on a well-proven model. It is the text used in advance to brief the technical team on the Waterton project on how a forthcoming drop-in would run and help to make them a little less anxious about meeting people. Essentially, it just tells a story.

Mrs. Smith heard about the event because the organisation she leads is on a stakeholder database. The invitation mentioned coming to the drop-in herself but also encouraging her members to do so, which she has. The invitation also asked her to make herself known to reception because there would be a special opportunity to personally meet the project team. (She knows that some of her friends also received a leaflet about the event and she herself saw the advert in the paper.)

She calls in on Friday afternoon. On entering she is asked to sign in; that way, she is told, she'll get a short report about the results of the event and advance notice of later events. She signs in and places a sticky dot on a map of the area, helping the organisers see where people have come from. She points out that she was asked to make herself known, so is immediately introduced to one of the project team.

She is given a short handout explaining what's on. There are a few things for her to do. Firstly, to visit some wall-sheets giving background information about the project. Secondly, to move on to a series of 'stalls' where she and others can add issues, check surveys and offer ideas and suggestions on all sorts of topics. The team member suggests she does this first and then meet up again for a one-to-one discussion.

She starts to wander around. She's particularly concerned about traffic issues so she goes to that stall first. On the display screen are big sheets listing the sorts of issues and ideas that have already been identified as important, plus a couple of explanatory maps. People are being encouraged to tick an issue or idea if they agree with it, elaborate the point if necessary and note what they think is missing. She spends ten minutes at this, adding ticks and comments. Others have been before her so the sheet is already getting full.

She notices comments books around for those who prefer to make notes in their own way, plus some maps on which people have suggested things to look after – major old buildings or key views. She also notices a laptop and screen in a corner where there is a live Twitter feed of comments from local people.

She then moves on to other stalls covering other issues and she adds more comments, ideas, ticks etc. The final stall and sheets draw attention to the plans as a whole. People have already commented on this, some querying the need for a plan, others suggesting wanting a plan in place quickly.

She leaves by late afternoon after chatting to other team members and adding a few more comments at a couple more stalls. On her way out she sees a sign she missed earlier about short workshops at set times during the drop-in to enable interested people to work with the team and get into more detail. She books herself into one the following day and makes a note to remind all her members (and her neighbours) to come along some time before it closes, or the following day.

Digging into this story reveals some of the basic and essential elements of a successful drop-in.

The event should be clearly *interactive*. This is relatively easy in the earlier stages when the action is about issue-raising, but similar exercises that are done at invited workshops about, for example, priority setting or option evaluation, can also be adapted for use in open events. As work proceeds, and options *plural* perhaps become one single option, details start to emerge and be almost fixed and that can make interactive events more difficult. Choosing the time for later events is therefore critical. They should happen *before* anything is completely finalised, while there are still some aspects on which people can comment and any changes, even if small, can still be made. If earlier events work towards agreed solutions and if there is a clear story of how the more final plans have built on earlier input, people are more likely to value a minimal final opportunity because they will be keen to see a plan finished or a project moving towards being built. At any later stage it is crucial to tell the story of how what is being shown draws from earlier work.[2]

Advertising is important. Existing local media, the press and radio are one bottom line. Plan/project websites are the next line of contact, especially if formally linked to other websites, for example, that of a local amenity group. Then there is the list of identified stakeholders and this is the time to work through them to invite all of any group's members along; perhaps accessing hundreds, even thousands of people. This is usually more effective than any invitation direct from the process manager or client.

Having some form of *reception* to welcome people is always important and it is valuable to try to make a note of everyone who passes through. As the story says, this is to be able to send them all a report of the event and to give them advance notice of the next events, which helps them understand that this is a process not a one-off. Most people will do this but don't press too hard. Getting them to add a sticker on a map to show where they live or work is also valuable and can operate as a gentle 'ice-breaker'. It provides an instant picture of who is/is not attending (just neighbours, just people from the north, only some from miles away etc.), enabling more targeted promotion next time.

Some people put on drop-ins staffed by someone junior or secretarial, there just to keep an eye on things. This just about enables minimal interaction, however, having *professional team members* present is immensely valuable. It enables explanations that can never be put on any short, sharp display boards and the many conversations bring the event closer to a collaborative workshop. This way, human dimensions and developing relationships between team and community can start to play a part.

Having a basic *handout* and clear, simple *background material* are both essential. They provide a starting point for people and save a lot of team time later (even if it is remarkable how many people don't bother to read material either).

Some people do not like to fill in Post-It notes or tick boxes on wall-sheets so it is important to give them a different opportunity. *Comments books* should be available with blank paper on which people can write comments (from experience, some are very short, others are long essays). Note also that some people may not like ticking wall-sheets because they are sensitive about their writing or spelling, in fact, some people may not be able to read. If that seems to be the case, team members can help, but with great care and sensitivity, always checking that what is written is what the person actually wishes to say.

Introducing something like a *Twitter feed* is becoming increasingly important. It does not just have to be what people are entering elsewhere, remotely; it can be added to by people in the room. There is a caution, however, because, if too well promoted, it can reduce the numbers attending.

Finally, it is absolutely essential to find a way to *report back* to those who attend (if they write clearly at reception, and not all do). Most of those who sign in will probably only want to receive a short summary of what took place and the results. The summary should, however, note that a full version of the results is also available, either on request by email, via the plan/project website or as a hard copy (see Chapter 7 for more detail on reports).

From events to sessions[3]

Using the word 'sessions' narrows down the coverage of this section because sessions are almost always something used in managed workshops rather than more open and inevitably informal, less timetabled events. That is not always true so this section also briefly suggests how workshop-type formats can be used even in events such as drop-ins.

Having gone through the process of planning the overall flow of sessions within any event, the next task is that of designing at the next level down. In other words, thinking through exactly what people will be doing during each session, for example, the 45 minutes allocated for addressing 'Issues and Aims' in Waterton (see box on page 93).

This section provides the main steps to follow in order to arrive at the appropriate design for any given session, a thinking process quite similar to that used for planning a whole event. Detailed guidance, for example, on how to group people for a specific task, is covered in Chapter 6.

Step 1 Decide on the purpose and outputs: These should already be known but digging into session detail may require going back to and perhaps amending them (so, once again, the iterative principle applies).

Step 2 Consider the human and practical aspects: These include:

- The **amount of time** for the session.
- The **space available**: Quantity, form, layout etc.
- The **energy levels of participants**: For example, is it after lunch, in which case can they do something which will keep them alert?
- The **confidence, feelings and knowledge of the participants**: What kind of activities/settings will make people feel more comfortable or less comfortable?
- The **confidence and skills of the facilitation team**: Don't try to do things that may be way beyond anybody's current level of skill or confidence.

Step 3 Consider who needs to work with whom: Should people with similar expertise or points of view work together or be mixed with others with different backgrounds, roles or opinions? Does everyone need to cover the same ground and do the same tasks, or could people be divided into groups, working in parallel on different tasks? This can be varied during an event.

Step 4 Frame the question/task: This is the most important single aspect of session design. People will not come up with clear, focused answers if they are asked muddled or inappropriate questions. It is therefore essential that any question/task is:

- clear and specific
- easy to understand
- not leading or biased towards certain outcomes.

Picking the right words to frame a question and task can take a while. Testing it with others can be valuable, especially if it appears complicated. Having decided the most suitable way of framing a question, think of at least one more way of explaining it so that, if necessary, it can be clarified. Examples can be useful to explain the kind of answers or responses being sought, but care is needed to ensure that this does not overly influence participants' responses.

Step 5 Design the session: In the light of what has been considered so far, now decide the best way to enable people to work on the question/task, i.e. how to allow time for:

- setting the scene or context for the task
- briefing the what and how of completing the task
- actually doing the task
- recording outputs

- sharing outputs
- closing or linking to any next session.

This list is important because it shows how much of a session might be taken up with things other than actually completing the main task.

Some form of recording is usually essential, although some exercises/tasks are about loosening up, awareness-raising or getting to know each other and may not need to be recorded.

When thinking about a question/task and in briefing people for it, it is crucial to make clear how any notes or results should be recorded. There are various ways to record notes and results (see Chapter 6) but they should be shared and agreed by everyone in a working group and in a way that is clear and open to all.

Step 6 Reality check: Once there is a draft design for the session, stand back and take a hard look at it. How will participants feel going through this session? Is the possible method/technique as simple as it can be (see next section)? Are the chosen outcomes and outputs realistic for the time available, mix of people etc.? At this point too, it is instructive and often imperative to 'multiply up', for example:

- Each group takes two minutes to feed back their results.
- There are six table groups.
- That could take 15 minutes for all groups to finish their feedback.
- Add a bit of slack and time to start and finish and the answer may be 20 minutes.
- Allow 10/15 minutes for discussion, if wanted.

As an example of all this combined, here is one format from the Waterton project workshop outlined in the box on p. 93:

> One session in the Waterton workshops was called 'From Early Opportunities to Initial Approaches'. The aim was for people to work in groups to suggest possible elements of a regeneration framework, drawing on their knowledge and what had emerged in the previous session.
>
> - One hour was allocated for the task.
> - The introduction was assumed to take ten minutes because it needed to link into key results from the earlier session and because maps and other materials had to be handed round.
> - There were to be three outputs from each group: An annotated plan, notes listing the various sites, projects or actions they felt were necessary and an overall rationale.

> - The first two were explained carefully but participants were told that what was needed for the rationale would be explained later. This helps to keep varied groups to time, avoids over-complicated initial briefings and, in this particular context, it was also thought that producing a rationale would be easier once people had done a plan and some notes. (See Chapter 6 for more on the dynamic of sessions, briefing skills etc.)
> - After 35/40 minutes the groups were stopped and a fuller explanation given of the 'rationale'.
> - People were then given a further 5/10 minutes.
> - All six groups finished very nearly on time (some slack had been allowed for this in the private programme).

It has already been suggested that work in-depth can link to work in-breadth by running small workshop sessions at a drop-in event. These seem to operate best over 60–90 minutes with perhaps 12–20 people. They are best advertised in advance but also on the day, something that someone on reception should direct people to. Rather obviously, any such workshop requires team input, probably also a facilitator. Such sessions just cover the same things and in the same way as in the main drop-in material but in more detail and with discussion and agreed responses. Alternatively, they can move things on to, for example, layout and options. In other words this can become genuinely collaborative work.

From sessions to methods

To (yet again) repeat a point, there are almost endless methods and anybody experienced in engagement will have used a wide variety, no doubt adapting them for their own situations and according to personal styles or even perhaps designing their own. Most of what follows covers methods for workshops but a few ideas are included later for drop-ins. (There is no Waterton project box in this section because almost all of the methods that follow were used, as were others, so the list would be too long to be useful.)

Some important provisos

There are several provisos to raise before describing some of the basic methods.
 First, continuing the cooking analogy:

- *Even the best ingredient can be ruined by using it in a poor recipe.* There is no such thing as a good method in isolation. Though some are patently of no

or limited value, any potentially good method only achieves that potential when used properly in a session or event and as part of a planned, overall process.
- *No recipe has a single ingredient.* There is also no such thing as *a* good method; any complex situation needs a range of methods to maximise access and to diversify the nature of feedback.
- *Adjust recipes and ingredients to available kitchen space, equipment, time etc.* You must use methods, do not let methods use you. Avoid adjusting a well-designed process to make it fit a specific method. Be really clear what needs to be achieved and that should lead to method choice, not the other way around.
- *Your cabbage, chocolate and rhubarb may be good but can you really bring them together into a recipe?* Make sure all methods used, and especially their results, can be used together to create a coherent, cumulative and advancing picture.

Second, all of these points lead to an important warning. As suggested earlier, there are some people who promote one specific method as the sole answer to everything. To continue the food analogy, they offer a 'fast food' approach, i.e. buy this method and all your problems will be solved, and usually in one quick event! Perhaps this warning is not necessary at this stage of the book but quick and apparently cheap fixes can be very attractive to clients, even communities. Which is not to say that they are always quick or even cheap; most are neither in the end.

The third proviso refers back to the earlier comment that too many methods are about the often easier early stages of trying to draw out information, issues and ideas. Fewer exist for the middle stage of processes and fewer again for reaching agreement.

And one final proviso. After only a very short time of practising engagement people start to develop their own methods related to their own situations and preferences. Look back at the Waterton example in the box ob p. 94 and notice that the session did not really use a 'method'; participants were simply given some resources (maps, large paper and pens) and asked to come up with some conclusions. It is crucial to be fully aware of some basic methods but also aware of when the simplest approach is better than some sort of clever trick or variation.

As only a few methods can be covered here, examples for all stages are included. Most are particularly appropriate in some form of collaborative workshop but some can be adapted for open, public events. All the methods are also content-neutral, i.e. they are not about any specific outcome (plan, proposal, project etc.) or topic (waste, housing development etc.).

Visioning

Methods for this have developed a rather negative reputation, mainly because visioning has too often been done in a cosy way that generates no more than 'motherhood and apple pie' outcomes. Many visions for plans in particular end up as equally applicable to anywhere or, in effect, nowhere. They have also too often been unrealistic, which may seem to be a counterintuitive comment about visions (are they supposed to be realistic?) but few situations start totally afresh and have totally free scope to explore, so there is always, and should be, a tinge of realism. The challenge is how to manage this without being too limiting because visioning is an important activity.

The method suggested here, called 'A Day in the Life', was developed by this author and has been used successfully in many different settings. Its success comes from it being grounded in day-to-day realities and because it starts from the idea that there may well be different, even potentially conflicting, visions of the future; what may be good for John (16) may not be so good for Mary (74).

Two bits of preparation are needed, both agreed upon by the client, ideally also by some potential participants:

- Agreeing a series of themes, for example, housing, education, leisure, transport, environment or safety.
- Generating short pen-pictures of a few local people and their lives *today*. These should be fairly negative and likely to raise issues for those people about the chosen themes.

The following is a typical household description from a session about various themes within the overall topic of transport:

> Dawn is a young, single mother. She rents a poor quality flat over a shop on the outskirts of town. She has to get her little boy's nappy changed quickly this morning because their bus to where she works, and where her son goes to nursery (ten miles away), leaves in only ten minutes. It was raining last night, and she has to walk along a lane without a pavement. The lane is used as a rat-run, and she knows exactly where she's going to get a splashing from passing traffic!

Participants work in small groups (of 4/5) and each group is allocated a different household to consider. An example brief was then as follows:

> It is now the year 2022 and your characters have been miraculously transported into the future (i.e. they are all the same ages as now). Working as a group, describe 'A Day in the Life' of that person or household that not

only reverses the more negative things in the description but also touches on as many as possible of the themes we are addressing today. We suggest you do it as follows:

- Make a long list, with everybody contributing, of all the things you can think of that would make a better life for your household.
- Check back through the list, to be sure you have touched on everything in the characters' descriptions and all the themes. If not, add a few more ideas.
- Work out how to summarise this on one sheet of large paper in terms of 'A Day in the Life' of your household; perhaps as a story or chart, a clock, even a picture – be creative!
- Put your character or household's name at the top of a fresh sheet of paper and start. (And remember that everybody else will need to come round, look at and make sense of your results.)

In the feedback, each group visits all other group tables and they note, where they can, a few things (i.e. suggestions for a better future) that are the *same* as suggested by their own group and a few things that perhaps *conflict* with their own suggestions (i.e. that might not be good for their own group's character). Shared discussion starts with the potential conflicts then moves on to draw out the ideas and features common to the different 'days'. Finally, discussion can move on to suggest some of the projects, plans or actions necessary to put those features in place.

This particular method has uses beyond just workshops. It has been done as a community competition in which people write their description for *themselves* in their town or neighbourhood a number of years ahead. For a regeneration framework in Salisbury relevant stories from local competitors, each a truly grounded but aspirational vision, were used to preface chapters in the final professional report.

Brainstorming[4]

This is a familiar method and is an easy and quick way to draw out lots of often interesting, challenging or creative ideas. However, doing it properly involves careful use of some simple ground rules, best introduced in verbal briefing but also noted on a handout brief. They are usually expressed as in this brief, again using transport as an example:

> We would like your group to brainstorm a range of possible actions, initiatives or projects to help make progress on improving city-wide transport, being as specific as you possibly can. Please think of things that

could be done at different levels (city, neighbourhood, even region) and by groups/organisations and individuals. Think widely; the list needs to include anything and everything: sensible, pushy, quick, easy, tough.

One person should write on the sheets in front of you (big pen, big writing for all to see please). Remember the key rules of brainstorming:

- No discussion.
- No objections.
- Everything is written down.
- More is better than (what someone might think is) better.
- Elaborate those suggested as well as offer new ones.

This usually generates *fewer items* than the method that follows because, as a shared activity, it often leads people onto things that, for them, are new and challenging and (despite the ground rules) this always generates some discussion. However, because of the cumulative or sequential nature of the task, the suggestions can be *more coherent*. But beware of domination, often not deliberate, by the holder of the pen.

Mapping moveable notes (Post-Its)[5]

The alternative to brainstorming as a small group activity is to individualise it, at least in the first stage. The focus of the task is likely to be the same as with brainstorming, i.e. generating issues, ideas or solutions, sources of funding, key barriers and so forth.

Verbal briefing is best, partly because it will be necessary to call time and move people on fairly quickly. The instructions (and be sure to have all the materials out) are basically as follows:

- Please each take a few Post-Its and write on them all the issues (or whatever) that you personally think this project might raise; just one issue for each note.
- Do this on your own for just a few minutes then we will move you on.
- Be sure to just put one issue per note and also only use the large pen (to keep comments short and clear to others).

Once it is clear that there are enough Post-It notes to fill large sheets of paper and/or people appear to be running out of ideas, call their attention, ask them to stop and then say:

- Now place all the notes very quickly on your large sheet of paper, this time working as a group.

- You should see immediately that some issues are identical, so place those notes close together on the sheet.
- Some may not be identical but are similar or related; move the similar ones into clusters.
- Keep moving the notes around until you have some clear groups of similar or related issues.
- Don't worry if there are a few 'lonely' notes; that often happens and they can sometimes highlight key things.
- Finally, draw a ring round any main groups or clusters and, if you can, give each group a name.

An example of the result might be as follows:

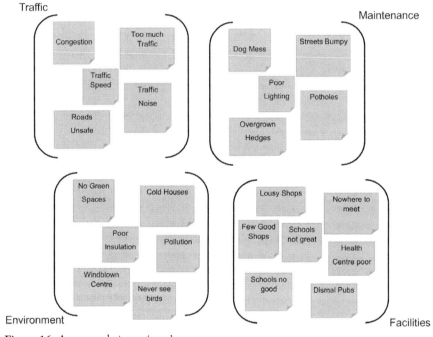

Figure 16 An example issues 'map'

As suggested, the results may be more *varied* than with brainstorming and it can lead to analysis through asking people to cluster the Post-It notes, thus generating broad *themes*.

Mind maps

Mind mapping (Buzan, 2010) is another form of generation activity which offers some things different to the two previous methods. It involves:

- using key words to record and develop the thinking of an individual or group on a 'spider diagram';
- highlighting connections between different elements, making it easier to make sense of a lot of ideas;
- creating a highly visual record of everyone's contribution and the way those contributions connect or build on each other;
- avoiding any hierarchy (nothing, comes out on top); and
- encouraging creativity.

This method is best done by a facilitator with a small or even large group, when the diagram can cover a whole wall. If done by small groups on their own the instructions need to be extremely clear. The result might look like this:

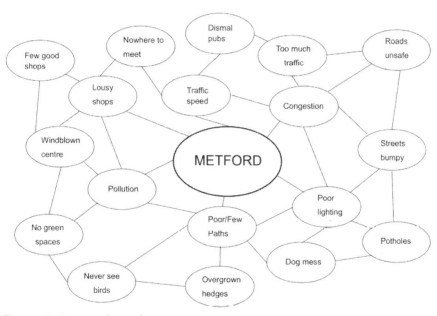

Figure 17 An example mindmap

- To produce the result shown in Figure 17 (using a fictional place at the centre), the facilitator started by drawing a central ring and writing in it the focus of the session: Metford.
- One person suggested something leading from this; that was written on a Post-It note and placed at the end of a line from the central ring.
- The next person suggested another issue/factor. That was added where that person said they wished it to go (on the same line, on a branch, on a new line etc.). That produced a new main issue and new line.
- That prompted a further suggestion which then prompted a discussion of whether this was a new theme/line or a link between some existing items.

- It was accepted that if new lines were added, links could be made back to previous lines.
- The next person added their issue and slowly, person-by-person, going round again for any repeat additions, the diagram developed.

To maximise involvement, each person can come forward to write their own addition (or add their own note) to the mindmap where they wish. Others might suggest the locations for additions but the person standing makes the final choice. If Post-It notes are used for items, and only faint lines are used for links, this creates scope to move the suggestions around and reshape the diagram as a whole.

This method is a more *visual* version of brainstorming (although the same ground rules should be used) and, by not being a list, it serves to better illustrate how items *relate* to each other. Although it may not bring out the range of personal points that one can get with mapping Post-It notes, it is extremely effective in encouraging an open form of cumulative idea-raising that can also generate overall themes. Final outcomes are best photographed, so large writing should be encouraged.

(The differences between these three methods in terms of group dynamics, quantity of results, quality of results and so forth highlights the proviso about the care needed in selecting methods.)

SWOT

The acronym SWOT is likely to be familiar to many. It stands for Strengths, Weaknesses, Opportunities and Threats. It is best used when it is likely to generate lists on all four topics, but be aware that one person's 'threat' may be another's 'opportunity'. The divisions never work perfectly but they stretch the agenda very usefully. It is easy to manage; the following is an example brief (about transport again) to some small groups:

- Please start by dividing your group's big sheet of paper into four quarters using the big pen.
- In the first/top left area note down what those in your group think are the current key **Strengths** of transport in/around the city (and give the box that title).
- In the second/top right area note what you think are the current key **Weaknesses**.
- Looking ahead, note in the third/bottom left area some of the **Opportunities** that might be seized to make for better transport.
- Finally, in the remaining area note down any **Threats** that might prevent it getting better.

- Just spend a few minutes on each. It does not matter where you start with but don't get stuck on one area; better to go round a couple of times.

If necessary, the ground rules for brainstorming can be used. The SWOT method can also be used in a large group. This can, however, become a little repetitive if all groups tackle all four SWOT headings, which leads to a major variation in which each group does one only then shares around, as we shall see.

Carousel

This is the first of two methods for moving from separate small group work to agreed summaries while avoiding a long and often tedious plenary feedback session.[6] These two methods allow people, by and large, to summarise for themselves.

Using the basic SWOT method, taking transport as the theme, you would start by asking each group (with a brief much as described in the SWOT section) to tackle just *one* of the SWOT questions. (For now this also assumes just four small groups.) They do, however, need to be told that other groups are tackling the other themes and that, having completed work on their one theme, they will be moving round to work briefly on all other themes. Because the sharing stage involves short, sharp stages and synchronised movement, it is best briefed verbally.

Once people have finished notes on their one topic, the facilitator might say the following (with management notes in brackets):

- You are now going to move around to have an opportunity to look at and add to what others have done.
- Please decide one person in your group to stay at your table.
- (Wait and check.)
- Can those remaining at their tables please draw a line under the notes just done?
- (Wait and check.)
- All others please stand but wait until I have finished saying what to do next.
- While those staying at tables remain seated, group 1 please go to table 2, group 2 to table 3, group 3 to table 4 and group 4 to table 1.
- (Wait and check.)
- Now you are there, please sit down and the person at your new table will take you quickly through what was noted. You are invited to add to (but not change) their lists in any way you wish. Either you can write the additions or the person at the table can write them.
- (After a due period of time ...)

- Can all those moving round please stand and move to the next table while the person staying behind draws another line under those extra notes?
- (Wait, check and if necessary direct to the correct table)

This goes through another two rounds, after which people will have added to notes on all themes before returning to their original group. Not surprisingly, each stage usually takes a little less time because there is (probably) less to add. Each group then looks quickly through what others have added and this then forms, in effect, the *whole* group's lists on all four SWOT issues.

This method can be used on other sets of issues or questions that are divided into three, four or maybe five headings (six is probably one too many) and, if necessary, the whole group can move round instead of leaving someone sitting there all the time. If someone stays, they miss out on the circulation, but having them there makes it easier for each arriving group to get started.

Cross-cutting

This is a variation on the Carousel method. Cross-cutting enables several aspects to be covered in parallel and it ends with a summary that is even more the group's own. Once again, having the SWOT example completed by four groups is useful.

In round one, each group takes four flip chart sheets, one each headed Strengths, Weaknesses and so on. The groups do exactly as in the basic SWOT process, i.e. make notes on all four aspects, but this time responses to each aspect are noted on the separate sheets. The facilitator then says:

- Can one person in each group please pick up your notes on the Strengths, one the notes on Weaknesses etc?
- All those with notes on Strengths to this table please, with your notes (pointing to the table/area).
- All those with notes on Weaknesses to this table with your notes (then the same for Opportunities, then Threats).
- Your new group now has four sheets all on one topic. Please take a fresh sheet of flip chart paper and work as a group to produce a summary of what you see to be the most commonly mentioned points on all sets of notes.
- This is not just about repeating all four sets of notes; it is about trying to find common points, a genuine summary.

As with carousel, this requires care in briefing and in managing the moving around, so is again a verbally briefed session. Varied numbers of themes and groups can be used but there should always be the same number of groups as themes.

Priority-setting, ranking and weighting

(We are now getting into the middle stage, the 'groan zone' territory where priorities are set and initial choices are being made.)

This adaptation of a method sometimes called 'Nominal Group Technique' or simply NGT[7] helps a group to move from having long, unrelated, unordered lists to some initial (with a stress on '*initial*') sense of ordering between items listed. That ordering may be about:

- **Priority-setting** or **Short-listing**: X is more important than Y.
- **Ranking**: Sorting all from most important to least important.
- **Weighting**: A combination of priorities and ranking.

Each of these will be described in detail.

The example – central area parking – is one that emerges quite often from work on the overall theme of transport. The specific example on which the detail here is based was the problems caused in the town centre of Stratford-upon-Avon by all forms of parking, for all sorts of vehicles and affecting all sorts of people and groups (cyclists, pedestrians etc.). A parking strategy was to be produced and some visioning work, issue generation and a SWOT analysis had all been done. The result was a long and rather random list of possible actions, lacking any sense of priority for which might best be taken forward. Here is how some **priorities** were generated.

Imagine the chart shown in Figure 18 as one or more big sheet(s) on a wall in a workshop. The left-hand column lists some of the suggested actions, one per box.

More car parks	
More pedestrian crossings	
Park and Ride by rail station	
Improved signage	
Coach park	
More cycle lanes	
Restrict delivery times	
Disability access throughout	
Free resident parking	
Better bus service	
Parking on Cattle Market site	
Information for coach drivers	

Figure 18 A priority-setting grid

The verbal brief to present to the groups was:

- We are now going to go through an exercise that we hope will result in some overall priorities among the items listed.
- Can I first check that everybody is clear what each of these items refers to?
- (Check, clarify, rephrase on the sheet if necessary.)
- You are each now being given eight sticky dots/stickers. Please reflect briefly on the list and decide which would be your own personal priorities or those of the group you represent.
- Once you have decided, come to the front and place one dot/sticker only in each of the boxes to the right against your eight chosen priorities.
- Please be sure to only place one dot/sticker per box and remember that the results of this are not final; they are just an interim indication of shared views.

In Stratford-upon-Avon, the outcome of this milling around (by simply adding up the dots) was eight actions that attracted dots from all or, almost all, present: the priorities or short-list. The emphasis on initial results is important because some people's priorities might not include any of those with the most dots/stickers. That and other points emerged from a short, and noted, discussion.

A richer but more challenging approach, which leads more quickly to a discussion about who chose what, is to ask people to respond to *all* suggestions. Using the same wall-sheets the facilitator says the following:

- We are now going to go through an exercise that will show which actions have full support or very wide support, and those which some might not be able to support.
- Can I first check that everybody is clear what each of these items refers to?
- (Check, clarify, rephrase on the sheet if necessary.)
- Can each of you please take a large pen and come up, in your own time, and place either a tick, a cross or a question mark in the small box on the right, for each possible solution.
- Use a tick if you personally, or you think those you represent, would support the suggestion or could at least live with it.
- Use a cross if that idea is definitely not supportable.
- Use a question mark if you are not sure about support or perhaps what exactly it might mean if adopted.

In another real-life example (the Romsey Future Plan), early work had produced a list of over 70 possible actions across various themes. As a result, this exercise was done over an extended refreshment break to allow a large number of people to access what was a large number of sheets. The outcome was 28 actions receiving all ticks and no crosses, 17 of which did not even have a question mark, and which could therefore be said to be agreed (though this was checked with the group). Several actions had many ticks but also some question marks. Further group work looked at those in order to agree (or not) whether or how the queries might be addressed. Though some actions attracted several (some many) crosses, none were removed at that stage because they might be considered again later if useful to support other agreed actions.

In order to show **ranking** between items, i.e. what people judge to be first, second etc., the adaptation is simple and again uses charts with actions listed, although this would now necessitate a final column of boxes to the right on each chart. This works best with a list of perhaps 12 items because people are asked to stay seated and consider their own rankings on all items (so 70+ would be an impossible task!). Having thought through their priorities, people come to the front and mark their rankings in the second column with numbers, i.e. 1 in the box by their first choice, 2 in the box by their second choice etc. Scores can then be added up and the total noted in the final column. The *top ranked* item is the one with the *lowest score* and so forth down to the highest score/lowest ranked.

Weighting is about not just priority but also relative importance. Someone's first three priorities might be, for them, far more important than any that follow, so a simple 1 to 12 ranking would miss the issue of the *weight* they attribute to items. To assess weightings everybody again receives sticky dots but this time up to perhaps 1.5 times as many as there are items on the list (10 items would equal 15 dots/stickers). They are then told that they can place *any number of dots/stickers where they wish*. That way someone might place four or five dots against their top item, while others get one dot each. Once again this requires the third column into which total scores are noted.

As for priorities, both ranking and weighting exercise require some final overall discussion of results (and results should still be **initial**).

Strategy grids

As things approach Kaner's third, narrowing-down stage, different and often more challenging methods are needed. There is no fundamental reason why people cannot take on the difficulties posed by such methods, especially because they are likely to have become familiar with both process and content through all their earlier work as previously discussed. The method that follows may appear quite complicated, but this reflects the fact that things probably *will* be somewhat complex by this stage. At the same time, it is striking how quickly people grasp the strategy grids method and quickly get under way, as long as everything is prepared in advance and the exercise is carefully briefed.

This method takes people a long way into moving from lists and priorities and into deciding on a sequence of actions, a programme or an overall strategy. The method does not generate a full and final answer (that must be made clear) but it serves to set up a group that is better able to comment valuably on a full version of a strategy once produced.

The Stratford-upon-Avon parking example is again used to explain the basic method but it can be adapted to many different situations where a number of ideas/solutions have been agreed and it is now necessary to decide what to do when and how the various ideas might connect together.

The advance preparation tasks are:

- Establish some themes for the strategy. In the real-life case the group decided on five:

  1. Costs and controls (e.g. parking charges per hour).
  2. Controlling access (e.g. service access only at set times).
  3. New provision (e.g. a new car park).
  4. Alternative modes (e.g. shift from car to bicycle).
  5. Information/education (e.g. better signposting to car parks).

- Write each possible idea/action/solution on a small card as per the examples in Figure 19 and also note the theme to which the action relates. Each workshop group requires one full set of cards. At least one spare set is useful as are some blank cards in case people wish to introduce a new or previously low-priority idea.
- Prepare a grid, as shown in Figure 20, in which the themes appear as headings to the rows and key timeframes appear as headings to the columns. Timeframes are determined by the nature of the task, in this case typical delivery periods for road and traffic initiatives. Each grid should be presented on an A1 size sheet and each group needs one sheet (spare sheets are again valuable).

Wider coverage by local shuttle bus Alternative modes	**Improved pedestrian routes into centre** Alternative modes

Figure 19 Strategy grid cards

Time Periods Main Elements	Short Term	Medium Term	Long Term
New provision			
Costs and controls			
Controlling access			
Alternative modes			
Information and Education			

Figure 20 Blank strategy grid

To be ready for the session on the day, for each working group lay out:

- a grid
- a set of cards, including a few blanks
- a glue-stick
- a pencil
- several sheets of flip chart paper
- a small marker pen
- a couple of large marker pens.

To get started, careful verbal briefing is necessary, supported by a written handout version. For this example, the verbal (and written) briefing was as follows:

- Your task is to devise a strategy for how, over time, all your agreed priority actions might best be implemented. Please look first at the big grid.
- The rows on the grid are titled with your own agreed themes. The columns show the usual timescales for implementation. Now look through the set of cards we have given you.
- These cards list all of your currently agreed actions and, in each case, the theme to which they relate. Please place all the cards in the very left hand box on the row with their theme title.
- (Wait until cards are located.)

- There are also some blank cards because further actions or ideas may emerge or you may wish to go back to an earlier idea not now listed.
- To get started, take some of the cards in any one of the rows (your choice) and spread them out along that row according to whether you think (just for now) that they might be short, medium or long-term actions.
- Now do the same with another row and perhaps another, without looking at the whole picture yet.
- Now start looking across and between rows because some actions might need to happen at the same time, so those cards should be exactly in line vertically down the grid.
- Those actions that happen/will happen just after or just before others need to be placed appropriately on the sheet, i.e. just before or after.
- You will find that some things are linked in some way across the grid; if so, either imagine a line between them or draw the line (but only in pencil for now).
- Some actions may start immediately and go on over time, in which case place the card at the start (left) and draw a long horizontal line, but only lightly, with the pencil.
- Some work may need to start now but the action does not happen till later, in which case place the card to the right and draw a pencil line back to when preparatory work might start.
- Avoid sticking anything down yet or drawing lines with a pen because you will find yourselves going round and round, moving cards several times, adjusting sequences, finding connections between actions.
- You do not have to use all the cards and can add other ideas using the blanks.
- Before starting, nominate one person in your group to take notes. That person's task (which should not stop them also contributing to the main task) is to try to note down all the discussion points, reasons and arguments used, especially why each card is where it is.
- We've allowed 50 minutes for this. Don't rush and we'll interrupt at some point to talk about how to bring it all together, for example, stick the cards down.

Now the groups can start, but it will almost certainly be necessary to circulate around the room, clarify and encourage people and then remind them again to be really accurate with locating cards and not using the glue-stick or large marker pen too early. The announced second stage reminders cover the following (not a full script here):

- Be sure all the cards are located exactly where you wish.
- Check sequences, links and possible lines.

- When you are happy, glue down the cards and draw the linking lines in pen.
- Reflect for a few minutes with your note-taker to ensure good notes to explain why the result is as it is.

This might take another 10/15 minutes. A worked example of the typical output follows in Figure 21, without all cards shown.

By this time participants will be ready for a shared discussion. Because the results will be quite complex, it is best to ask the groups to walk around to look at all results, to note where there appears to be some common agreement about what to include, about when certain solutions are going to be implemented and where there is clear disagreement. This quickly clears the top line of agreed actions and timeframes, and the bottom line of those needing further debate. In the real-life example, some groups were asked to do a minimum action strategy, some a middle range strategy and some a radical action strategy. This helped to promote vigorous discussion and raised a powerful point about the value of consensus building. At the outset of the parking work, several stakeholders were extremely negative, wanting nothing to be done. However, when it came to talking through the three different approaches in the grid session, the whole group almost refused to talk about the minimal action version because they now all agreed that something quite dramatic was needed!

(As hinted earlier, there are far less methods for the later stages of consensus building although one key source is Friend and Hickling, 2004.)

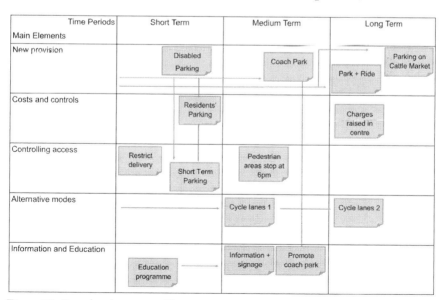

Figure 21 Completed strategy grid

Securing final agreement

At some point there will be a need for final closure and, if at all possible, full agreement to the plan, strategy or project. However, it is common for some stakeholders to be challenged by, or even baulk at, finally agreeing. Because there are occasions when full agreement is not possible or not judged to be necessary, it can be valuable to agree with any group what might constitute *appropriate* agreement if *full* agreement cannot be reached. This is best warned about ahead of any last stage rather than at what could be a tense time towards the very end.

Although this approach is included here as a method, it is not that simple. The example that follows, again about parking, uses a range of approaches; in some cases built in earlier in the process and in others using the methods introduced above. Some key points to bear in mind when developing one's own approach are added at the end.

Following the strategy grids session, officers used the results to develop a preferred option, using technical information about procedural lead times, delivery periods and, importantly, project costs. The option presented was clearly based on the middle range strategy that had emerged, with some additions from the radical group's work, notably some high profile, short-term actions.

At the final workshop, all previous stages, information and conclusions were revisited and checked. This led to suggestions that some actions that ranked low in the shortlisting session should now be included because they would complement and reinforce those in the emerging strategy. This exercise also highlighted aspects within the proposed strategy on which a few stakeholders were unconvinced and hence unsupportive. Short group work sessions added some important changes that converted the non-support to support, if sometimes still conditionally.

Having cleared away much of the uncertainty, stakeholders were asked whether they could support all aspects of the strategy in general but still with some cautions or not at all (or not enough for it to proceed). This was done through a combination of a wall-sheet on which they marked their conclusion against each aspect and commentary forms, mainly for people to note any cautions.

Because of the careful work to that point, no stakeholder refused to support the draft strategy at all. Of the 30+ people present, all but a few supported it in full and, once again (this repetition is crucial), there was an opportunity to work a little more on those issues that were still unresolved for the small number of those still uncertain. This was not entirely successful and a few (very few) remaining concerns were formally noted. That formal noting was all that was necessary to enable the remaining few people to then agree to support or, as you may recall, say that 'we can live with', the now slightly revised final strategy.

This led to another interesting outcome about consensus processes. The strategy was announced to local journalists at a media event. Being well used to ignoring statements from the top table, the journalists anticipated going straight to others for negative comments. They were totally confused because there was no top table; all the key stakeholders shared in making the announcement because everybody had agreed!

The key points about final resolution are therefore as follows:

- Ahead of time, agree on the principles for reaching agreement or handling disagreement.
- Reconfirm these principles at the start of any final stage.
- Recap all previous work, revisiting if necessary to prevent them surfacing again.
- Do not assume that full agreement will come the first time the question is asked.
- Be prepared to go back to look at, and try to resolve outstanding cautions or queries.
- People can agree to an overall result while still having points of dissent: 'I can live with that'.
- If there is any significant dissent, this needs to be formally recorded.
- Involve all parties in sharing and communicating the outcome to others.

Drop-in methods

The aforementioned methods were developed mainly for use with groups working collaboratively. Some of these, as well as the others that will be noted, can be used in open, public drop-ins such that, although people contribute as individuals without discussion, there can be a form of cumulative, group result. (And to close the circle, some of the methods that follow can also be used in a workshop.)

Issues sheets: In the early stages of any project it is important to discover, rather than assume or guess, people's issues or concerns. Lists can be done by the technical team but, with some local networking or via an opening workshop, lists can be produced that can genuinely said to be from local people. That makes a real difference in terms of responses; it 'rings bells' more quickly and it shows that the core team is sensitive and listening.

Figure 22 shows a version of one of six issues sheets used at the first drop-in about a housing development in Modbury in Devon. The large sheets of paper were put up on a wall. People were given large marker pens and asked to place a tick against any issue they agreed with, a cross against any they disagreed with, to elaborate anything on the list (by using the item's number) and adding other issues on the deliberately empty bottom part of the sheet (which other people could then tick).

Issues and Ideas: ACCESS AND MOVEMENT

1. Access to the site to be from the A379 (Church Street).
2. Safe pedestrian and cycle links from the site into the town centre.
3. Great care and careful consideration of the Palm Cross Green junction especially in regard to pedestrian safety.
4. Safety on the approach, at the entrance and around the site.
5. Pedestrian priority across the site, with slow speeds for car movement.
6. Footpath routes around the site………
7. …….and if possible to the wider countryside.
8. Good street lighting.
9. Safe pedestrian routes to the school.
10. Links to open spaces and play area for residents of both the new development and existing areas of Modbury.

Figure 22 Example issues sheet

The sheets can fill up quickly so spare sheets and blank paper can be valuable. However, any sheets taken down should be kept on display somewhere to show the overall results.

Maps with notes: An effective method not just to get people active but also to get them talking to each other, even in a drop-in, is to have a central table with a very large map on it of the plan or project's area, town, neighbourhood or site. (A huge aerial photo is even better but that can be expensive.)

People can write notes and either place them directly onto the map to make a point – 'great view here', 'difficult road junction there' etc. – or place them around the edge and draw a line in to the specific place they are commenting on. Once the map starts to fill up, which can happen very quickly, it can be replaced, putting the full one where people can see it. Without even asking (but asking does help), people naturally start to put ticks, crosses and comments on other people's notes. One can also have different colour notes for positive, negative and neutral (i.e. purely descriptive) comments.

What makes (a place) special? Local people are often quite protective about their place and believe it is special and distinctive, even if they also recognise that it needs improvement. This can be drawn upon by asking people, for example: *'What makes Waterton different from the nearby towns of Springfarm, Westnewton or Spaylin?. How do you know you are in Waterton and not those other places?'* People write their ideas onto a large sheet or, once again, on Post-It notes; the latter providing the freedom to move the notes around and group them as an event goes on.

Scoring scales: Instead of just ticking or crossing, people can be given a more evaluative option by using scales. Having generated some key issues, the scale can offer perhaps five choices for how people feel about that issue or idea, as in the example in Figure 23. The scale can also be used for people to score what they feel to be the importance of each of a set of issues.

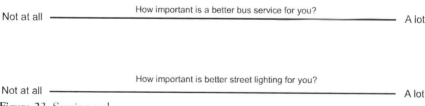

Figure 23 Scoring scales

Key messages

- There is no substitute for very careful overall event and session design. It pays off hugely and everything will suffer if the planning has not been done with great care.
- Avoid just planning for the supposedly 'core' activities. All the small things such as briefing, walking around and moving paper take real time.
- Time out (e.g. for refreshments) is positively valuable, certainly not lost time.
- It is always crucial to allow some slack in any programme.
- Be sure to make a conscious and careful choice of methods for the specific context, adapting strictly as necessary.
- Make sure that results from all methods can be added together to generate a rich and coherent picture.
- Finally, the importance of good face-to-face work should now be obvious, so do nothing until you have read the next chapter.

Notes

1. See Chapter 6.
2. A classic approach is to include a series of 'you said so we did ...' notes.
3. Much of what is in this chapter is adapted from material worked on by this author and others credited in Chapter 1. However, it is important to note that this material draws in particular from the (unpublished) work of Andrew Acland and Rowena Harris.
4. The term 'brainstorming' is occasionally queried because of possible links to problems of mental illness. This has been checked with NGOs working in the field of mental illness; they have no concerns.
5. Some people call this method 'metaplan' but the term is almost useless in non-professional or mixed settings.
6. Chapter 6 introduces some ways to get maximum benefit in minimum time from plenary feedback sessions.
7. As with the need to avoid the term 'metaplan' when working with non-professionals, it is good to avoid the formal terms.

References

Buzan, A. (2010) *The Mind Map Book: Unlock Your Creativity, Boost Your Memory, Change Your Life*. Harlow: BBC Active.
Friend, J. and Hickling, A. (2004) *Planning Under Pressure*. London: Butterworth-Heinemann.

6
Working with People

Introduction

After all the preparation, things can still go seriously astray once people are, as it were, in the room. Without denying the importance of careful preparation, the true test of collaborative working is how well all those involved work together to produce sustainable solutions. Enabling people to work together creatively is a fundamental skill for achieving agreed approaches or solutions. That requires some form of intervention or management commonly called facilitation. However, the common interpretation of facilitation as active, constant management belies a whole range of other less direct, sometimes quite hidden, ways in which people can be enabled to work together well. Facilitation also only succeeds if all other aspects are in place; it cannot compensate for poor stakeholder selection or a bad process.

One definition of facilitation is 'to make easy',[1] which suggests that successful dialogues or events are those where, on reflection, participants wonder why the facilitator was there at all. A delightful illustration of this came from an eight year old. The mother of this child was on one of this author's facilitation training courses in Italy, near the end of which she was asked to help to facilitate a workshop. Her husband was unexpectedly going to be late home so she took along her eight year old daughter. The workshop went well, all very active, with small groups, plenary sessions and clear agreement around key issues. On the way home, the mother asked her daughter, who knew she had been on some sort of course, what she thought facilitators do. The girl pondered a moment, then said, "They move chairs and collect cups"!

Rather oddly this is a sign of real success because, occasionally, a workshop appears to virtually run itself and all the facilitator seems to do, or is seen to do, is to set groups going, keep to time, share results, seek agreement ... and collect cups. Unsurprisingly, if that happens smoothly, it is a result partly of the care that went into the preparation but also of some hidden work by the facilitator at the event itself.

More often than not, however, some aspect of an event requires intervention, for example, because of uncertainty about the scope or differences between participants. Just occasionally, truly significant issues arise requiring very robust, highly principled and strong intervention. The aim of facilitation is to always aim for the most minimal role while watching for, and being ready to act promptly on, any potentially damaging issues of content, process or people that may arise. In a sense, a good facilitator is a still point at the centre of a whirl of people, issues, ideas and results; it may be a gentle whirl, it may be enough to blow people off their feet. Yet without some degree of 'whirl' no progress is made.

In addition, as mentioned in the opening chapter, facilitation is not something to be turned on and off, a trick or a device. Treat it as such and, though nobody knows how it happens, people can almost 'smell' that a task is being done solely because it has to, and that will affect everything. Put the other way round, people will be aware, even in difficult circumstances, of when a facilitator is genuinely doing their best on everyone's behalf, and that can help to build trust, even if difficult issues still remain.

Before any reader stops here because they do not personally wish to become a facilitator, it is important to point out that much of what is in this chapter is about basic skills which will be of value to anybody who finds themselves working with others and wishing to get the most from such situations. Learning to listen, to respond with comments that take a conversation forward, or to settle someone who is uncomfortable or being overly challenging, are all valuable skills in everyday situations, be those one-to-one exchanges, chairing a meeting or even just being a participant in an event attended and facilitated by others.

Yet this leads to a dilemma for this book because there is really only one way to learn properly about facilitation and that is through active, practical training and then by practice. Nevertheless, beginning any training course (see Appendix 4) with some knowledge of the terms, aims, settings, approaches and repertoire of key skills, as outlined here, can speed the training and enhance its value.

As with the previous chapter, this chapter starts with some quite broad issues and then becomes more specific. It covers:

- **Why facilitation?**: Introducing those situations that can be damaging if not managed (facilitated) properly.
- **Facilitation is and is not …**: Clarifying how facilitation differs from, for example, arbitration.
- **Setting the scene**: Highlighting important things to consider even before an event starts.
- **The basics of facilitation**: Covering what can get in the way of people working well together and how in general facilitation can help.
- **Key skills**: Introducing details of some key basic skills.

- **Support techniques**: Covering techniques that can help to back up and support the key skills.
- **But there can still be problems**: Focusing on some aspects of how to deal with particularly challenging situations.

Although original authorship is difficult to establish, this is the chapter that draws most from work done by those credited at the end of Chapter 1. Some specific credits are given where material draws particularly heavily from one person's material.

There is only one Waterton example in this chapter because the list would otherwise be endless. However, everything that follows proved necessary, probably crucial, in the generally successful delivery of the Waterton project. Offering references is also difficult because the chapter draws from a whole medley of sources, some not even published, and some material that has been adapted by many people over many years. Three books listed in the overall references are however of real value, those by Mann, 2014, Australia, 1997 and especially Kaner, 2014. (The latter goes into far more detail than it is possible to cover here and it also covers aspects about event planning as in the previous chapter.)

Why facilitation?

An exercise used regularly at the start of this author's training courses asks participants to recall any experiences of failed meetings or workshops and to catalogue the reasons behind the failure. Some of the things often listed as going wrong have been mentioned in previous chapters but warrant repetition because of what they tell us about people and their behaviour. Here are just a few of the most commonly quoted and, for some readers probably chillingly familiar, reasons:

- **Venue and timing**: Holding a three hour interactive event at 5.30pm on a Friday evening in a room with fixed seating, awful acoustics, one flip chart stand and no refreshments.
- **Unclear scope before**: People arriving at an event unclear exactly what it is about and what they can influence through their involvement.
- **Unclear scope during**: Lack of clarity before an event, worsened by a lack of clear explanation at the start.
- **'It's all a fix'**: Assumptions made that any engagement or involvement is a superficial, cosmetic exercise and the outcome has already been decided.
- **Professional language**: Introductions, exhibition material, task briefs etc. littered with professional jargon.

- **Gender/age differences:**[2] Far too often (still) middle aged men dominate discussions and the input of women, young people and those from minority groups is undervalued or not valued at all; even worse if the men are the professionals.
- **Unclear facilitation**: No planning of group sizes or membership, no clear leadership, unclear instructions, tasks taking too long, no or very boring feedback and no clear outcome or obvious next stage.
- **Bringing the baggage**: Someone arrives with and constantly refers to negative experiences of past situations, regardless of whether those apply again or not.
- **Grandstanding**: Someone insists on standing up and giving a lengthy speech, probably about issues other than those being addressed.
- **Distrust of facilitators**: A conviction that the facilitator is there solely to get what the client wants or is 'in the pocket' of key players.
- **No feedback**: No feedback is given to all those who attend, nobody knows where any results have gone or how the results have or have not been used.

There are no miracle solutions to these problems, and others not listed, but properly designed processes, careful stakeholder identification and invitations, clear information, good practical planning and, in particular, effective facilitation can recognise, pre-empt and address them.

Facilitation is and is not …

It is extremely important that the term 'facilitator' is properly understood and distinguished from others.

A facilitator is *not* the same as a **chairperson**. A chairperson is part of the group, openly a stakeholder, entitled to express opinions and may be the final decision-maker, so is in no way independent. A chairperson can, however, act in a facilitative way, for example, by ensuring that everyone has a real opportunity to speak or by keeping the meeting on course.

A facilitator is *not* an **arbitrator**. An arbitrator is usually a content expert, engaged by some parties, usually two, to listen to evidence and then make her/his judgement, to which the parties will probably have agreed to abide. This is therefore an active role in terms of content so this has no place in facilitation work.

A **mediator** is closer to a facilitator because he/she does not offer comment on the content of the matters at hand, focusing instead on enabling the parties to more effectively negotiate their own agreement. Mediators usually work with two or three parties who are in dispute and mediation is particularly valuable in high conflict settings.

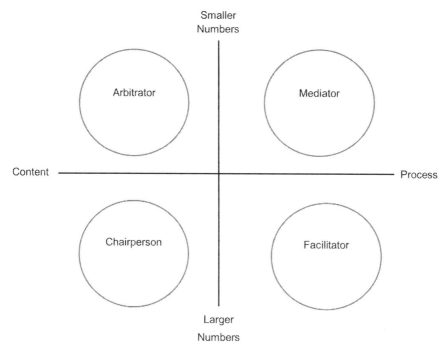

Figure 24 Chairperson, arbitrator, mediator, facilitator

There are few differences about how a **facilitator** and a mediator operate except that facilitators tend to work with larger groups and more actively manage and shape the discussions. Neither facilitators nor mediators are part of the group they are working with and their role is focused solely on the process not the content.

The main differences, summed up in the diagram shown in Figure 24, highlight our concern here with the right side and mainly with the bottom right quadrant.

Setting the scene

Before moving to outline core skills, it is important to highlight some practical points. If 'facilitate' means 'to make easy' then everything that follows in this chapter is about that. If, for example, people arrive at an event and everything is ready, if they are received warmly, given all necessary information, directed to their own clearly marked table and shown where they can get refreshments, they will already be on their way to focusing on what they are there for. All of this attention to creating a positive, relaxed but still focused 'climate' or 'feel' in a room gives people confidence and creates trust in any organiser or facilitator because it clearly shows respect and courtesy.

Although what follows relates to some form of workshop where people choose, or have been chosen, to come together, almost all of the points and the skills also apply, if to a different degree, to many other events such as drop-ins. This is mainly about what was described earlier as a good facilitator's hidden work.

There is no right answer for the first point, **timing**, because no one time can ever be right for everyone. Midweek evenings can be problematic for elderly or young people, especially in winter and when transport is limited. Evenings can also be challenging if many invitees have already experienced a full day's work. A midweek daytime event has the opposite benefits and disadvantages. It is likely to be light outside, public transport is more regular and those for whom the session is part of their work can attend more easily. However, those unable to leave work, for example, a shopkeeper, are disadvantaged. Shifting to a weekend is different again. This can be a more equitable draw (if bigger) on people's spare time but again people such as shopkeepers would be disadvantaged if a Saturday was chosen. Choosing either a Saturday or a Sunday can have negative implications for particular faith communities. (Friday evenings have this particular problem but are best avoided anyway.)

Running two sessions, afternoon *and* evening, can maximise attendance but also keep key groups (notably professional and lay people) apart when they ought to be discussing together. If there is a *series* of events, there are arguments in favour of altering timings (more can attend at least once) and arguments against (groups may be different at different events). If there is any solution at all it is to focus on careful stakeholder selection, giving priority to those considered to be most important and who might be judged to be least flexible. Contacting people and agreeing a best time with them can be effective and creates a positive start. In addition, having a coherent overall process in place, as in previous chapters, allows other opportunities to engage more and different people should some not (or not be able to) attend any one specific event.

One additional caution: Once a group meets for a first time, it is common to agree the next date with those present. This may be practical but, by definition, those who could not make that first date will not be present to agree a next date that suits them!

The second point is about the **venue**, and related issues. If the focus is on group work, a venue with rows of fixed seating is inappropriate. Having or creating a large, flat space is always best. For longer workshops, say over a full day, there can be value in providing break-out rooms for groups but, because of how long it takes to move people around and the loss of focus that often follows, it is only of value if group work sessions are of at least 45 minutes. Disability access, audio loops etc. are absolutely basic requirements.

If each group has its own facilitator, groups of 12 are just about manageable, 8–10 optimal. If groups are self-managing (see later) a maximum size would be seven people, groups of five or six work well but variety and stimulation can be lost with only three or four. (A practical tip is that most venue managers quote room sizes in terms of numbers seated in rows. Divide that number by three if there is to be group working.)

The quality of any room is also important. Too formal and this can inhibit open discussion, too informal and this can inhibit focused discussion. Natural light for daytime events is very important, although good and maybe additional lighting for any presentation is also important. If there is to be a presentation using a projector, that requires laptop, table, screen, blackout curtains/blinds and probably extension leads.

Almost all workshops must also have some form of 'wallspace' so as to be able to put up flip chart sheets, blank pieces of paper for moveable notes, maps, group work results etc. Wallspace can literally be walls but portable screens are usually more flexible, although they have to be robust if people are to be writing things on sheets. Beyond that, creativity is often needed to create wallspace; a well-known trick is to stand a few tables on their ends to make a screen! Be sure to check in advance what type of fixing is appropriate (tape, pins, Blu-Tack etc.).

The size of group tables should relate to the size of any groups formed. Smaller tables are generally better because they can be combined if larger groups are also needed. The big, square or round tables often used by conference space providers or hotels are too large for group work; people can neither hear nor see each other properly across a large table. Small, lightweight tables are most adaptable, not least because workshops often involve shifts from small group working to plenary sessions. The same applies to chairs; the lighter and more easily moved the better.

Arrangement of any tables in the room is also important in helping to create a collaborative feel. Facilitators often refer to the preferred arrangement as 'cabaret style' as illustrated in Figure 25.

Workshops do not often involve equipment other than that necessary for some presentations. Many venues provide flip chart stands, although these can be of little value because of the importance (see later) of putting *all* notes, instructions, results etc. up on walls or screens; a flip chart stand can only show one sheet at a time.

With time and venue decided, the next aspect is **refreshments**. Clearly, there is a very bottom line of providing access to water and then drinks (tea, coffee etc.). That would probably suffice for a morning, afternoon or evening event, unless some people are travelling a distance or coming straight from work, in which case they would need some form of food before or after (as appropriate). For any full day event, the provision of food should be considered essential. This is not just a matter or courtesy; people enjoying food and drink, bumping into new people and talking to each other in a non-formal setting is in its own right an important and genuine part of creating a constructive atmosphere. (Also see the section 'Arrivals exercises' later.)

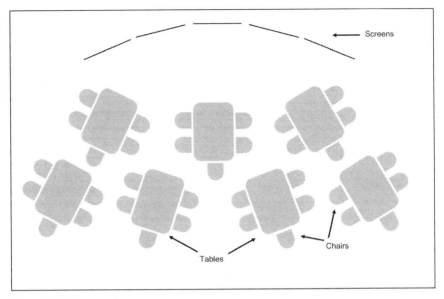

Figure 25 Cabaret style room arrangement

The final practical aspect is the **setting up** on the day. Arriving one and perhaps even two hours ahead of the start is essential. Tables, chairs, wallspace, equipment all need to be laid out. Tables can be numbered with stand-up group numbers or letters. Spare tables for handouts, prepared sheets, maps etc. are important and everything should be laid out ready in event sequence. Badges for participants can be valuable, whether formal (printed in advance) or informal (self-written on envelope labels) and be sure to also have badges, in similar style, for the facilitation and technical teams. Arrangements need to be planned for any refreshments. A reception table is usual and will require someone (or more than one) to staff it, copies of any handouts, participants' lists, badges and so forth (and spares of anything sent in advance because some people forget). If groups are pre-arranged this can either be notified on participants' lists or badges or with lists on tables, the latter being more likely to get them to the right table. Or all of these, just to play safe.

The basics of facilitation

What can get in the way for people?

Even with all practical points addressed, certain things can still make it difficult for people to work together and reach agreement. Some of the key things that get in the way are described here, some of which relate back to the earlier list of what can go wrong:

- **Taking positions**: People may start by taking and holding a position, believing that this is their best defence against the views of others with whom they may disagree.
- **Compromise/consensus**: People have varying views or experience of what is possible and, in particular, some have a firm belief that consensus is unachievable and/or negative views of compromise.
- **Uncertainty**: Some people may feel uncertain about aspects of a situation, their role in it or ability to contribute, and remember that *some* things in *all* situations are likely to be uncertain for *all*.
- **Complexity**: Any issue, problem or challenge that is in any way interesting will contain many separate, often conflicting factors to which different people will attribute different weight or value.[3]
- **Secrecy**: As with position-taking, some people 'play their cards close to their chest' and are cautious about revealing their concerns, information or ideas, for fear that others will belittle them or steal them.
- **Power**: Power inevitably varies between people in superficially different groups, professions or sectors, but it can happen just as readily within what seems to be a like-minded group and can also vary on different aspects of a situation.
- **The past**: As suggested, this can create baggage which gets in the way of progress, yet past experience can also offer useful precedents and avoid 'reinvention of the wheel'.
- **Time**: There is never *enough* time, however, timescales set by professionals too often fail to reflect the way in which and speed at which non-professionals think and work, especially when they are working with others they have not met or worked with before and addressing new issues.
- **Resources**: As with time, there is never enough, especially when this is considered to just mean money.
- **Cultural factors**: This is often thought of as factors such as class, gender, ethnicity, religion or sexual orientation which affect people's comfort in certain situations, for example, in terms of personal distances, voice levels and social hierarchies, but be aware that professions, local authorities and private companies also have their own cultural patterns and preferences.

Any or all of these factors can get in the way of progress and a facilitator needs not just to be aware of such things but also ready and able to tackle them if they appear to be having an effect. At the same time, everybody, even the apparently most difficult person, brings something that can help to move an issue forward and generate a creative solution, be that information, judgement, experience or ideas. While operating in a way that deals openly and, if necessary, assertively with complexity, uncertainty, cultural factors etc., the task of a facilitator is to enable people to focus on the positives and to help them realise and use the positive things that they bring.

Some of the principles or standards through which a facilitator can achieve this are:

- **Equality**: Ensuring all voices are heard and respected.
- **Responsibility**: Taking the full leadership role required of a facilitator.
- **Cooperation**: Encouraging people to work well together.
- **Honesty**: Making clear and dealing with all key points or issues.
- **Transparency**: Being completely open on all content and processes.
- **Accountability**: Clearly being seen to be on no one side.
- **Focusing on process**: Avoiding any intervention in the content of any discussion.
- **Independence**: This is not an absolute (see later).

The key things a facilitator can do to help this are to:

- Create a shared focus or purpose.
- Ensure everyone has a chance to contribute.
- Shape managed discussion.
- Ensure that facts are communicated clearly.
- Keep facts separate from opinions.
- Help people to understand each other better.
- Encourage some new reality or solution to emerge.
- Show that the event/session purpose has been achieved.

This then leads to the main skills needed by a facilitator, some but not all of which will be elaborated later:

- **Listening**: Listening actively, patently paying attention and being sensitive to the emotion behind what is being said.
- **Empathising**: Being able to see what the other person sees, hear what they hear, feel what they feel, and letting them know that this has been done.
- **Clarifying**: Helping people to be clear about what they want.
- **Reframing**: Helping people to develop statements or points that take discussion forward, not stop it.
- **Questioning**: Asking the right person the right question in the right way at the right moment.
- **Affirming**: Letting people know that they and their opinions are valued.
- **Non-verbal communication**: Noticing and interpreting non-verbal communications and using one's body language to communicate.
- **Observing**: Noticing what is going on among stakeholders.
- **Event management**: Keeping control of any event and ensuring it achieves its objectives.

- **Time management**: Ensuring the time available is used effectively.
- **Adaptability**: Being able to react quickly and flexibly to changing circumstances and the needs of participants.

Key Skills

Listening[4]

Listening is an important skill for everybody but it is absolutely essential for a facilitator. Misunderstanding and conflict often begin with poor communication, especially if someone feels (or even knows) that what they have said has not been listened to properly. It is still too often necessary to have to help people to put aside old patterns and ensure that communication, especially listening, is significantly improved.

Most people are never listened to properly. The significance and benefit of simply giving someone full attention cannot be underestimated; it can often significantly change a person's behaviour. As a facilitator it is, however, not enough just to listen; it is important to *show* good listening by:

- Not allowing interruptions (or interrupting).
- Giving plenty (but not continual) eye contact.
- Relaxing, leaning forward, nodding one's head from time to time.
- Perhaps using encouraging sounds (but don't overdo this).
- In any spaces/gaps, following up by using questions to encourage them.

Few people (including facilitators) speak in easy to remember phrases or make bullet point type statements; they jump around and appear to ramble, especially when emotions are running high. In fact, even if it seems that just facts are being stated, emotions are almost always also being communicated. A facilitator needs to show respect and empathy and be aware of and then acknowledge any emotional undercurrent, listening for natural points of emphasis (voice volume or physical signals/gestures). This background dynamic needs to be attended to as much as the main content of what is said.

If someone 'explodes', avoid trying to stop or control it immediately; wait a little and then, when it's over, acknowledge the strength of feeling before calmly asking the person to summarise their main points for all others. This also helps to depersonalise comments if the explosion is aimed at someone else in the room.

Finally, if someone's comments are to be recorded (for example, on wall-sheets visible to all) try if possible to use the same words that have been used. If it is necessary to summarise, check any summary with whoever spoke.

As well as being good practice for recording, this helps to communicate to everyone the importance the facilitator attaches to everybody's contributions and it demonstrates good listening.

Questioning[5]

Almost invariably, at some point someone will make a statement. If not intercepted by the facilitator, another person will almost certainly respond with a counter-statement and so on into the canyon of conflict. The way to intervene in this is to shift to asking questions because questions generate answers and offer a way forward. Asking the right question in the right way at the right time is at the heart of successful facilitation. Good questioning can start to bring structure and clarity to the apparent complexity of an issue and enable people to move on by separating:

- fact from feelings;
- personalities from concerns;
- manageable and specific issues from all-encompassing issues;
- needs and interests from positions.

Key points for a facilitator about any form of questioning include (sometimes with examples):

- Before asking a question, think about its likely effect on the person.
- Never ask a question without really wanting an answer and without being able to live with any answer given.
- Make sure any question is based on a genuine need to know, not a statement or criticism in disguise.
- Warn people before asking a tricky question to show that a sensitive issue has been recognised. (*'I guess this is a difficult question, but I have to ask ...'*)
- Questioning can rightly be challenging but avoid any feeling of interrogation by explaining why it is necessary to challenge. (*'I just want to ensure that we've explored this properly.'*)
- After asking a question, wait, don't rush. Silence should be interpreted as important thinking time, especially for people unused to questioning that is aimed at moving things on.
- Remember the importance of tone of voice and body language.

In advancing a conversation, it is natural (and particularly so for professionals) to want to move quickly and to get to, and hopefully subsequently clear away, any negative issues. However, successful progress depends on first establishing some kind of rapport or *common ground* with someone. It is only once there is something positive on which to agree, however minor, that someone will feel comfortable about addressing tougher issues. Getting to that point is usually outlined as a series of three types of questions used sequentially over three stages:

Stage 1 Reflective question: This is what it sounds like; it is a 'bounce-back' to the person who has made a statement, using what they have said to get them to (ideally) agree, to say yes. Someone may say: *'It's a great idea ... but the council are bound to reject it!'*

At this point, resist the temptation to ask why they think the council are bound to reject it, i.e. do not challenge the *negative* element of the statement. Instead, pick up on the first and *positive* part of the statement by asking: *'So you think it's a good idea?'*

The answer could hardly be anything but *'yes'*. And, if they are only saying *'it's a great idea'* for effect but don't mean it, better to know there and then.

Stage 2 Open question: This is designed to ensure that yes/no answers are all but impossible, encouraging the respondent to elaborate and provide more information. Once again there is a choice: Asking about the *'great idea'* (the positive) or asking them about why *'the council are bound to reject it'* (the negative). It is almost always better to go first for the more positive version by asking: *'Why do you think it's a good idea?'*

That should elicit some elaboration but beware because it will probably also prompt further comment on why it might be rejected!

Stage 3 Directive question: This type of question is designed to take matters forward towards a potential area of agreement or to clear away disagreement. It might even be a specific proposal. So, having heard what made it a great idea one could ask: *'So what can we do here to be sure it remains a good idea?'* and even add (before the person says it): *'... and minimise the chances of the council rejecting it?'*

So now, in this example, one has already moved into the difficult territory of clearing away the barriers in relation to the council, but rapport has been established and there are some very basic building blocks of positive progress in place.

Reframing[6]

Sometimes people say things (deliberately or by default) that worsen a situation or make it difficult for others with different views to listen. The facilitator can help to prevent this happening by acting as an intermediary and refocusing the statement, bringing out any positive potential or at least minimising damage. Typical responses that can cause concern include:

- **Over-particularising**: Getting into real detail to which others cannot respond.
- **Over-generalising**: Treating a specific situation as a global (and therefore insoluble) one.
- **Personalising**: Making a comment that is indirectly, sometimes directly, about someone in the group.
- **Jumping the gun**: As in *'It is obvious, the answer is to ...'*.
- **Diverting**: Introducing a clear 'red herring' issue.
- **Looking to the past**: Either *'it worked fine last time so don't change it'* or *'it failed last time so let's not make the same mistake again'*.

Reframing or refocusing is a powerful way of managing such comments but great care is needed when refocusing statements. If done badly there can be accusations of being biased, manipulative or patronising, even insulting. So limit reframing to one of the following purposes:

- Clarifying information.
- Emphasising common goals.
- Confirming common ground.
- Eliminating abusive or extreme language.
- Clearly identifying underlying interests and needs.

Reframing can be done simply by changing the syntax of what is said, for example, by turning statements into questions, or by using less provocative language. Each of the following examples deals with a slightly different type of reframe (in bold). Each starts with a statement that needs reframing, followed by a possible facilitator response:

- **From 'your' or 'my' issue to 'our' issue**
  - Comment: *'Why is everybody so negative about any proposal?'*
  - Facilitator: *'Would it be useful to talk about how we can all evaluate proposals?'*
- **From specific to general**
  - Comment: *'I think we have to focus on the site for the new landfill.'*
  - Facilitator: *'Looking at it more broadly, can we start by talking about how the community's waste could best be dealt with?'*
- **From statements to questions**
  - Comment: *'The answer to traffic congestion is to build a bypass.'*

- Facilitator: 'So what are other people's views on how traffic congestion might be addressed?'
- **From single to multiple frames**
  - Comment: 'Who should be allowed to bring a car into town?'
  - Facilitator: 'Can we talk about different people's different transport needs?'
- **From personalisation to depersonalisation**
  - Comment: 'Peter has not been doing his share of the preparation work.'
  - Facilitator: 'So, how can we ensure that the preparation tasks are shared by all?'
- **From past to future**
  - Comment: 'Whose fault was it that the system failed last time?'
  - Facilitator: 'What can be learned about how things were done last time and how can we apply them here?'
- **From threat to affirmation**
  - Comment: 'Let's talk about the council's insensitivity to this concern.'
  - Facilitator: 'What is it that makes this so difficult for all concerned?'

Finally, don't be too clever. After trying a reframe, look for confirmation that it is acceptable to the participants and, if in any doubt, ask.

Support techniques

As well as the core issues of how a facilitator handles comments or questions to engender constructive conversations, other almost hidden things can help to either prevent awkwardness in the first place or help to make it easier to deal with.

Arrivals exercises

This is a very valuable way to create a good atmosphere or the right 'feel' in a room and can also generate useful results.

A traditional public meeting individualises people (though some 'hunt in packs'), giving the impression that the event 'belongs' solely to the organisers and hence establishing an 'us vs them' relationship from the start. Arrivals exercises used for workshops enable people to be active from the moment they arrive, meet and talk to new people, and start to establish a sense of a genuinely shared event.

An arrivals exercise can be as simple as everybody taking a marker pen as they arrive and ticking some boxes on a grid. It can involve asking people to suggest issues by filling in Post-It notes that are then grouped. People enjoy watching this and, as they do so, often chat to each other and also, very importantly, to the team. On one occasion where a high level of conflict was expected because a key issue (camping in the Avebury World Heritage Site) had been avoided for many years, the moveable note responses were requested in terms of *'what will happen in five years' time if this issue is not addressed and solved now?'*. The mapped result was so powerful that, even before people sat down for the formal start, it was clear to all that a solution *had* to be found.

Finally, if a really active arrivals exercise is used, those running the meeting may well have to ask people, perhaps very loudly, to stop what has already become 'their' meeting in order to start 'your' meeting! The change in atmosphere and feeling of meeting 'ownership' that this can create can be dramatic.

Ground rules

It is extremely important to establish ground rules at the very start of any session. If there is no explicit agreement about how to behave – what to say and not say, how to treat each other and so forth – someone will probably take advantage of that vacuum, deliberately or by default. From the opposite direction, people always prefer to know the format, style, boundaries or ground rules for any event. This is especially important when there are likely to be significant differences of age, gender, literacy, self-confidence, power etc. between group members. As far as possible, these factors need to be equalised and up-front agreement about how to do that is a key factor in preventing inappropriate behaviour.

A common example of how important ground rules are is seen when people try to insist on making speeches. By agreeing on the ground rule of 'no speeches' at the outset, speech-making can be prevented from even starting. If someone then stands up and tries to give a speech, although it is the facilitator who will remind them of the agreed rule, the rule will work because the facilitator is clearly seen to be acting on a *whole group* agreement.

Ground rules can be developed with a group at the start of an event but they are relatively standard, so the usual solution is to prepare some, share them with the group (on an easily visible flip chart sheet), check for any suggested changes or additions and then confirm. This can be done quickly for a relatively non-contentious event but more care is needed, even checking formally that all agree, if a contentious event is anticipated.

There are no fully agreed lists but this one has proved its value:

To get the best out of the time together,'* it will be important that:

- all views are heard – that people listen as much as talk;

---

* This is often a more comfortable phrasing than the overly formal sounding 'ground rules'.

- everybody has an equal right to have their ideas and comments recorded;
- everybody shares their thoughts as fully as possible;
- people try to stick to time and task (as will we);
- this is not an occasion for long anecdotes or speeches;
- there is a shared commitment to go forward;
- full reports are made on each occasion – nothing is lost or edited.

According to the context, it may be appropriate to add one or more of the following:

- No swearing.
- Have respect for each others' views and differences.
- Be happy and have fun.
- Don't fear challenging opinions.
- Avoid jargon (or explain it).
- Avoid being judgemental.
- Tolerate people who are not great at spelling.

There can also be a need to encourage people to be open by making clear (perhaps just verbally) that:

- Nothing said or noted here will be attributed in the report to any named individual.
- That also applies outside this event to all involved – please do not attribute comments made here to any individual present.[7]

In recent years a very practical extra rule has been introduced: *'Please switch off your mobile phone or put them on silent'*!

Briefing

Participants at a workshop, as individuals or groups, will almost certainly need some briefing or instructions for a task. As a facilitator gains experience it can become easy to overlook how tasks and briefs can appear to people who may be fresh to interactive ways of working and even cautious about them. To avoid this, one should try to become a naive participant and almost literally 'walk through' any exercise or task, seeing everything from a participant's perspective. Alternatively, ask a colleague to read any briefing, task instructions or materials to see if they make sense. The ultimate aim is to rise above all the complexity and detail that may have gone into designing a task and make things as simple as possible for the participant.

Briefing can be done in one of the following ways or, which is usually better, through a combination:

- written or verbal;
- at the start or once underway;
- for a whole group, for small groups or for each individual.

Written instructions should be as short and sharp as possible and it is important to separate *what* people are being asked to do (and information that goes with it) from *how* they are to do it. A good way is to italicise one or the other or, if writing on a wall-sheet, to use coloured pens.

For handout briefs, use large type sizes and spread things out to make it all easier to digest quickly. If people need to refer to an information sheet, avoid putting that on the back of the brief or they will have to turn the sheet over and over. It can seem wasteful to hand every individual a briefing sheet but sometimes this is the only way to get it across, and occasionally somebody re-reads the brief aloud, helping to keep their group on track. If a brief is written on a wall-sheet, make sure everybody can see it clearly (so this is less good with large groups).

Verbal briefing is almost always essential. This may mean just reading through the written brief, using this as an opportunity to reinforce certain points, elaborate others or phrase things in a slightly different way to get the point across to someone who may not have grasped it.

Most briefing happens at the start of an exercise but this is often not enough, so expect to have to chase up, clarify or remind people. There can also be occasions where a completely full briefing at the start would be mind-numbing and instantly forgotten. People may get the basics quickly so that they can start and that can be followed up with further detail, even stages in a session, once they are on their way.

Finally, there can be occasions when it is valuable to intervene with further briefing, either when there is a need to move people from (for example) stage 1 to stage 2 in a task, or in order to inject further information.

The 'Strategy grids' method (see Chapter 5) illustrates almost all of this. The background and preparation are quite complex but that should not be conveyed to participants. Written briefings, even for just the first stage of this exercise, always require some follow-up (with an individual, a group or everybody) and the session goes through three clearly separate stages.

Managing group work

In most situations people working in small groups will be self-managing and do not require their own facilitator. The in-between approach, if more than one facilitator is present, is for each facilitator to stay close to and generally assist but not stay with, maybe three or four groups.

Different group dynamics are generated by the length of time to be spent on a task and the balance of membership:

- **Time**: Any first exercise should allow extra time, at least five minutes, for people simply to chat and get to know each other (whether about the task or not doesn't matter). At the other extreme, even with careful briefing, it is difficult for a group to keep focused for much over 45 minutes on a single task or stage.
- **Group membership**: There are various options for this. Self-selection is not generally advisable because people will usually join their friends, collude and narrow down rather than open up. Pre-allocation to groups before an event, mixing by role, geography, sector or whatever, is probably best but it needs to be explained. Occasionally people will object ('*I came with my friend*') and some gentle adjustment is then advisable (without opening the floodgates). Sensitivity is also needed regarding any cultural issues. For example, in some societies a group could contain both men and women, but in other cultures it can be essential that they are separated.

Care is needed on the day to ensure that all necessary materials are laid out ready, centrally or on each group table. And results need to be collected, usually at the end of each task unless needed for the next task (and see the section 'Recording' later).

To encourage groups to start after briefing, some rephrasing or repeating can be valuable (as shown) but avoid waiting for everybody to know absolutely everything. It is better to get most people started because the uncertain ones will either pick it up from others in their group or you can spend time with them alone to offer extra assistance.

Once under way, there are still things to be done:

- If the private programme allocates 15 minutes for a task, it is better to say '*10 minutes or so*'.
- If people have to do, for example, brainstorming then shortlisting of points to feed back, this is two stages, so warn them and allocate time accordingly.
- Once started, give occasional reminders if it is a longish (20+ minutes) task, and/or a five minute countdown to the end.
- Keep an eye on who is making notes and make sure that as much as possible is being noted in an appropriate format; help if necessary.
- Don't rush; some groups prefer to just talk and then make their notes in a late burst with a few minutes left.

- If it seems that a group has split, there is more than one discussion going on, one person is dominating or some seem to be left out, remind them promptly and gently; don't let it drift.
- If the issue is about a person, try to intervene in a (reframing) way so that one person is not in 'the spotlight'; talk about how the group as a whole is working.
- Similarly, if one group is struggling, make any intervention sound as if it applies to all groups not just one; for example, stand back from a slow group and announce to everyone to pick up the pace.
- Use any originally unannounced spare time with care, but start to clearly chase when only X minutes left.
- Gently up the pace to finish on time.

One benefit of breaking a task into stages is that this creates an opportunity to intervene, enabling any slower groups to catch up and creating a short breathing space which also helps the facilitator to keep to overall time. For staged tasks, encourage (don't force) the group to share the writing/recording tasks.

Be aware that finishing can be difficult for certain people. Some people who seem to be going quickest can be the most reluctant to finalise because new ideas or issues will suddenly arise. Some worry about finishing because it suggests complete agreement. Others can be worried if they are going to present, call out or leave their results for others to see. In all such cases, make clear that any results from their work are strictly interim and not in any way final.

Group work almost invariably produces a lot of paper. If possible, before the start of the session mark up all relevant sheets with group number/letter, task name and any specific group theme or focus. Then keep numbering as things go along or help self-managing groups to do this. Once a task is complete (and unless results need to be kept for further sessions), check that all results are properly noted and then collect everything, putting items in correct session order.

Managing feedback from groups

A completely understandable fear that people have about group work is the awful, boring, group-by-group feedback sessions that simply repeat everything; it is OK for the group that goes first, but a total waste of time for the group that goes last! Success with any other approach is as much about explaining it carefully, and early, as it is about actually doing it. So, in the overall event introduction (which is best), or perhaps as part of the first task briefing, say something like this:

- Most of your time will be spent in small groups.

- I will be drawing out and summarising only the key points during the feedback sessions.
- This is mainly so that everybody has some idea of what others have covered.
- Nothing that you have noted will be lost; the summary is less important than all the specific detail from each group.
- And all that detail will be in the full event report.

There are a number of short, sharp, clear and valuable ways to manage feedback. The classic approach is a form of key points call-out. If using this approach, be sure to tell people about it at the outset of any task. Then call a halt with a few minutes left, explain exactly how the feedback will be done and ask people to spend the last few minutes preparing for it (and, as before, allow for this in session timings).

Once everyone is ready, the call-out session might be briefed verbally and managed as per the following (using issues as the content):

- In this feedback session, we are not going to repeat everything that groups have done.
- We will be trying to capture key points that are common and perhaps different between groups.
- I will take just *one* issue from each group in turn and each following group then needs to offer something *different*.
- Please keep comments as short as you can – just a few words.
- Group A, what is your first key issue please?
- (Note this issue on a wall-sheet. Intervene promptly but gently if they are spending too long or offering more than one.)
- Group B please, what is your first issue that is *different*?
- (Note again, intervene etc.)
- Group C – a different issue please.
- (Note again, intervene etc.)

Continue the process around all groups. With just five or six groups, go round the same way twice or three times until people feel – check with them – that most of the key issues have been drawn out. With a lot of groups it is best to go from (for example) group A to group H then back from H to A, equalising the input so that group H are not always last. Be assertive about closing off when few new ideas are emerging, but stress again that this is because the result is just a *summary* of *key* points.

Once this is complete, ask for any overall reactions – *'are there some surprises, new combinations, unexpected thoughts etc?'* – and note answers. Avoid letting this go on too long or it can become a full repetition of group results. Close by repeating that it is the detailed group work that matters most and that every comment will be in the full event report alongside the summary.

This format not only speeds the process along and avoids boredom but it also helps people to note that other groups have said some of the things that their group has noted (which is always reassuring), to hear other, possibly quite different points and to get a feeling that the final, overall list is as comprehensive and widely agreed as it could be. This creates a real sense of shared understanding and is another way to begin to build common ground.

For tasks with several stages, an alternative approach is to start by, for example, asking group A to read out their main conclusions on stage 1. Other groups can then add anything important to this (additions to be noted) or raise queries. Group B then does the same for conclusions on stage 2 and so on. There are often fewer stages than groups so always ask for additions first from those who will not be presenting about a stage.

Recording

These are the skills, when writing on flip chart sheets, of capturing what can be complex arguments in bullet points in a way that satisfies the person making the argument. This is important for note-takers in small groups but especially important when recording, for example, a call-out as described earlier. This is very much a matter of practice, but here are some hints:

- Use large writing, even with small groups, to make the notes legible to everyone.
- Use all capitals or capitals and lower case, whichever is easiest, quickest and clearest.
- Try to start note-taking very quickly so as to become the clear focus of the discussion. This also helps to set the pace of a discussion, reassures participants and helps to clarify how a full record will be produced.
- Ideally, always record the exact words of a speaker but, if paraphrasing, check with the person making the comment.

Tracking

While people are focusing on what matters most, as they are supposed to do, they can worry if suggested timings change or if they lose track of what's been done and what is still to come. If this happens, potentially peripheral things (see 'Rhubarb sack' section) can 'knock them off track'. Tracking tackles this, is remarkably quick to do, almost subliminal and extremely valuable. It involves the facilitator keeping an eye on times, stages, coverage, materials, people etc. for two reasons:

- To ensure that things stay exactly (or near enough) as promised in any programme, or resolve how to proceed if things start to drift.
- To remind people of what is presumably a sensible sequence and process, where they have reached and why, and where they still have to get to.

Very simply, at regular intervals and without any special emphasis, people can be told where they have reached, whether changes have been made and why, whether this will affect overall programme, finish time, lunch etc. They rarely worry about changes but simply need to be reassured that someone else is keeping an eye on such things on their behalf.

Rhubarb sack

Any interesting issue has fuzzy edges and it can be easy to slip unnoticed into spending time on aspects that seem perfectly reasonable when raised but which are some way off the main focus and are in fact diversions at that moment.

Two examples during the Stratford-upon-Avon parking strategy process were changes that might happen in nearby towns (totally outside the scope and certainly not worth waiting for) and possible electrification of a rail line (clearly important but no announcement was due until well into the process).

The rhubarb sack[8] provides a way of registering and accepting the potential importance of certain issues without them getting in the way of the particular focus of a particular meeting at a particular time. In the two examples given, the two points raised at the first meeting were written on a wall-sheet, which was then displayed throughout all meetings. By the fourth meeting the electrification announcement had been made, so its implications could be discussed sensibly at that time and, at the last meeting, the group chose to come back to the issue of action in other competing towns.

Placing items in the 'sack'[9] reassures people that the issues are not being ignored but neither are they intruding, except when they ought to, and the group can make its own decisions about when, or whether, to return to them.

Recording decisions and actions

Action planning can be done in several ways. The version described here is about ways of reaching and recording decisions or future actions to follow any event. Using a chart similar to Table 4, any decisions are noted separately from any actions necessary to move the decision forward. This is because some decisions may lead to a number of potentially quite different actions (and different people to work on them). The chart also seeks the name(s) of who will take responsibility for each specific action and suggests, perhaps not too precisely, by when any action is expected to be taken.

Table 4 Action planning grid

Theme	Action	Who by	By when
Congestion	Journey times data	PB	Mid June
	Community views	JG	Early September
	Ideas from elsewhere	PT/AM/LT	Before summer
	Research charging	PB	Mid June

This is usually done with a large, flip-chart paper version of the chart on the event wall. It can be valuable to have the chart up throughout an event because decisions and hence actions can arise from the very start (e.g. a need for extra information). Having the chart up for everyone to see is important because, once several actions have been noted, it often becomes clear that starting one action is dependent on finishing another, so that would probably require changes to the 'by when' column, perhaps also to the 'who by' column.

Event reports

It is essential to produce some form of report of all meetings and events. A full report of a meeting (see following box for an example summary report) needs to be as full and verbatim as possible. There are several important reasons for this:

- Nobody will have edited the core content of the report. However well a professional, officer, or someone in power composes an edited version, participants will always distrust it. Transparency must be based on full and comprehensive, shared information, as free of bias as possible.
- As work proceeds, options begin to appear and choices narrow down, so people need to be able to track this for themselves. They value being able to refer back to exactly what was stated and recorded in the past in order to assess the legitimacy of later ideas or proposals.
- Once people realise that full reports are being made, they not only relax more but also become more concerned to ensure that their comments are recorded properly.

In order to respect these points and to avoid dictating detail, which can and should vary, there are two main ways of producing a report: Typed and photo reports.

Typed reports are the most common form. They are valuable for participants as a reminder but also essential for someone not present in order to help them understand what took place. Typing results also ensures that they can be stored electronically and (for example) lists of ideas can be grouped together and re-categorised, then assembled for future use.

There are almost ideological debates about whether to correct people's phrasing or spelling in reports, but the principle must be that the participant is always right, so nothing is changed unless essential for comprehension. Original material should always be stored (until a sensible period after process or project completion) so that participants can check back if queries arise.

Typed reports also have value because they can include linking and explanatory text, material contributed following an event (e.g. participant contacts or a message from someone unable to attend) and additional material, but all this must be clearly distinguishable from direct event results. The usual way to do this is to note all actual event results in plain text and all explanations or added material in italics. One short extract from a Waterton project workshop report illustrates this (note the plain and italicised text):

> *There was just enough time left by this point to do a brief summary call-out of key points from each group. People were asked to be ready to call out two key points from their group notes. Each group called out a different point in turn, going round twice. The key points raised were as follows:*
>
> - Make the river area more attractive – e.g. with the sluice.
> - Link the docks to the town centre.
> - Focus the town centre around the town bridge.
> - Develop a night-time economy.
> - Increase unit size and diversity amongst the shops.
> - Celebrate the many historic features.
> - It is not radical enough.
> - Make links to the various cultural nodes.
> - Regenerate St. Jude Street and Broadway.
> - Create a high quality market area on Overway.
> - Make it all a flagship destination to attract/retain people.

The simple and quick alternative approach is to create **photo reports**. At its quickest, and assuming that lunch is provided at the end of an event, one can photograph the notes/results, paste the photos into an electronic format, print them off there and then and hand the report to participants as they leave! At the minimum, there would normally, of course, also need to be a cover sheet with details of the event, notes about the programme and perhaps a participants' list; these can delay things unless ready in advance.

With a bit of practice, modern digital cameras or mobile phones are perfectly adequate to take usable photos of flip chart sheets, one sheet per photo. Four photos placed per page in a report works well if people have written large and legibly (otherwise one page per photo). This has two advantages over typed-up reports: There can be no tinkering at all, and photos can show diagrams and linkages across text that typed reports may find it difficult to show.

Full reports may be all that is necessary for workshop reports because those involved are likely to be keen to check detail. However, for more open, public events and even for some workshops, a **summary report** of one or two pages can also be valuable. This inevitably involves some editing so care is needed. Most importantly, any summary report (or the message that accompanies it) must also tell people where they can access the **full report** if they wish to check anything.

But there can still be problems ...

No matter how careful one is, life can get complicated. Facilitators are often warned that participants X and Y are certain to be 'really difficult to deal with', or 'you'll never get Z to contribute sensibly'. One facilitator's experience was rather more blunt: *'You'll never get people in this town to agree on anything!'*[10] Dealing with problems as they happen is a specialist territory and action learning is accepted as the only way to develop the relevant skills needed to deal with them. However, there are ways that can help to prevent problems, reduce the chances of them happening or limit the effects they may have. That is one area that *can* be covered here.

There are also real and regular difficulties experienced by facilitators who are members of staff of an organisation running a collaborative project because they are unlikely to be seen (at the outset) as genuinely or appropriately independent. How to tackle that is also addressed in the next section.

'Difficult' people?

Why is this title in inverted commas? It is because, in a way, there is no such thing as a difficult person and everybody (including you) has almost certainly been seen as 'difficult' at one point or another. Labels such as this can also be deeply damaging because, for example, the people who are labelled as 'difficult' can vary dramatically. From experience, some are extremely keen but just tend to go over the top and some are sceptical and non-committal. Both of these are potentially positive stances, although there are also some who have a habit of interrupting but only out of sheer frustration at having experienced so many badly run meetings.

There are, however, some people who set out (or appear to set out) to spoil, and that is genuinely negative. Only rarely is that about the facilitator or the specific issues being addressed; it is more usually brought in from other settings or previous histories as 'baggage' that needs to be dumped. If someone insists on offering you a warning about person X or Y, treat that warning with caution because, if the facilitator offers any glimmer of recognition of that difficulty during a session, it will only make it more likely that it will materialise; it will become self-fulfilling. In other words, follow good practice procedures and run the event exactly as it should be run; be alert but do not be diverted.

There are, nevertheless, several basic aspects of collaborative ways of working that make it more difficult for negative people to hold sway or less likely that they will be 'difficult' at all. The key methods to use (which are good practice anyway and several have already been mentioned) are as follows:

- **Selection of participants**: When it is clear that there has been a conscious and transparent process of selection of participants, this implicitly identifies and partly legitimises both the particular people and those they represent at an event. Sharing the list in advance adds to the value of this.
- **Advance briefing**: Good briefing in advance should make clear to people that they are part of a clearly designed and managed process, robust enough not to be sidetracked by disruptive comments. Briefing can also be explicit about the nature and character of an event, even spelling out the ground rules in advance. That challenges anyone who might seek to dominate and reassures those worried that they might not be heard properly.
- **Room and group arrangements**: Having a public meeting-style front platform and arranging an audience in rows leaves you with an event tailor-made for anonymous, individualised comments, objections, shouting, complaints etc.[11] Placing people into small groups, arranged informally around a room with no obvious front (and no platform), helps dramatically. It breaks down the 'us vs them' problem and binds people into feeling that any comment (or complaint) they make is directed as much at others in the group as at the event managers.
- **Facilitation**: The easiest (in fact almost the only) thing to do in a public meeting is to point to the platform and complain about what 'they' have done (or not done). As soon as someone is present as a more independent facilitator, it becomes more difficult to direct blame or responsibility to 'them'. Facilitated meetings are more neutrally run, even to the point that strong objections are as clearly noted as anything else.
- **Arrivals exercises**: Public meetings individualise people and establish an 'us vs them' relationship from the start so that the meeting clearly 'belongs' to the organisers. As explained earlier, arrivals exercises are very effective in changing the dynamic of 'ownership' of meetings from the very outset.
- **Ground rules**: This is perhaps the most direct way of addressing potential difficulties and has been covered in this chapter. Once everybody has

agreed to some rules there is something that must be broken; far better than leaving it all open.
- **Group work and summaries**: There has already been an emphasis on the value of active exercises involving whole groups, designed and managed so that everybody contributes. Placing the emphasis on group results rather than overall summaries is also of value because it lessens the chances of domination by a single awkward person.
- **Full reports**: There is nothing more chastening for someone who makes an inappropriate comment than to see that comment being recorded verbatim by the facilitator and then to realise that it will be printed for all to see in the full event report! If the completeness of event reports is made clear to people at the start, this will reduce the chances of someone acting up during an event.

'Independence' for non-independent facilitators

Once again inverted commas are used here. In this case it is because there never really can be an absolutely, genuinely independent facilitator accepted as such by all. Equally, just because someone is paid for their work by a client does not mean that they cannot be appropriately independent in that particular context. An appropriate level of independence is something that must be made public and transparently demonstrated (including to the paying client) over time.

Even external facilitators, clearly not part of a client organisation, can find it difficult to detach themselves once their names are associated with that client. Experienced facilitators will be familiar with the situation in which, even after managing a whole series of meetings as an outsider, someone still assumes that they are just a full-time employee of the client.

This suggests potential problems for any facilitator who is an employee of the initiating organisation, something that is almost certain to remain common given the small number of freelance facilitators. Problems can arise for an internal person because:

- They are likely to be known to some participants and to have engaged with them in traditional professional mode on previous occasions.
- That can lead to them being targeted with comments about previous initiatives, or challenged about long-standing grudges or niggles.
- They will probably have direct or indirect knowledge about a place, its people and the issues under discussion.
- They are quite explicitly 'one of them' and seen by many as a perfectly legitimate target for any sort of comment about any sort of issue, whether or not it is connected to the subject of any particular event.

Given how most initiatives have been run in the past, such responses are of course completely understandable. In order to tackle this, the key words (also relevant, if less so for external facilitators) are 'appropriate', 'transparent' and 'demonstrate'. There is now good experience that shows not only that in-house staff can achieve what participants later judge to be an appropriate level of independence, but also that their more detailed local knowledge and experience can be of positive value if used with great care.

So how can appropriate independence be demonstrated transparently? Much depends on individual style and attitude but several building blocks can be put in place and certain basic behaviours used. What follows is aimed at those people tasked with designing a process, facilitating events or simply helping with facilitation from within their own organisation.

Even before the start of a process there is value in taking the lead responsibility for any invitation letter or briefing information, i.e. to have the internal facilitator's name on it, not that of the technical leader. Any letter or briefing note can then be a first opportunity to point out that the facilitator, not the technical team, will be taking responsibility for the process. In addition, one can use this opportunity to avoid what are often highly formal or bland letters and notes; a facilitator can often inject more informal and engaging ways of writing. This helps to create the impression that whatever is being set up is not just another technical tick-box exercise.

During most events, and certainly at any first event, it is absolutely essential (if also sometimes very difficult) for the facilitator to stay completely out of any content discussions, even if wrong information is aired or certain key issues are left out. Always make sure that there is someone else in the room to whom any content questions or points can be directed. And then:

- At the very start, describe the situation openly, state clearly the intention to ensure that the meeting is equitable for everyone and encourage people to highlight any occasions on which they think this is not being done properly. In other words, be open and up-front from the start.
- Ensure the ground rules highlight a number of ways in which the meeting and its management will be equitable and independent.
- Be prepared to intervene assertively and promptly even if it is a technical colleague who oversteps the mark. In fact, the first time this is seen to be done often causes a loud intake of breath but it can significantly increase the group's confidence that the facilitation will be independent.
- Be explicit about how comments are recorded and particularly attentive to capturing people's exact words and phrases.
- If necessary, about halfway through, check back with participants; ask them straight out if they think the facilitation is being handled properly.

- At the end, make clear that you, as facilitator, will take prime responsibility for ensuring that the full report is indeed full and unedited by others.
- Chat to people after and get any informal feedback.

There are also some typical situations that arise, for example:

- A local facilitator is likely to know some of the participants, which makes it tempting to launch into conversation with them about other issues. Although it is essential to be welcoming and sociable, avoid such conversations if possible; better to chat with those one does *not* know.
- An issue can arise on a topic other than that under discussion, perhaps one the facilitator is dealing with in another setting. Avoid handling these within event sessions, make it clear that the other issue is different and talk about it during the break or at the end.
- At the outset, people may refer back to the previous ways of working within the facilitator's own organisation, even to their own personal behaviour (e.g. on the platform at a public meeting). Do not get drawn into discussing this, certainly not into explaining or justifying it. Assert clearly that what is under way now is different and that it should all be treated as a new start.
- However, people will understandably wish to know how this new situation might be different to and better than previous ones. There is no possible way to answer this at the outset, so don't pretend or waffle. Simply explain that the approach being taken is different, that it is based on work elsewhere that has made a genuine difference, and encourage them to stay with it and see. Make clear that it is not now (at the start) but only as the process unfolds that they can ever know if this approach is better.

Key messages

- This chapter cannot teach you to facilitate; that can only come through proper training and lots of practice.
- The ideal is to be as invisible as you can most or all of the time but clearly visible when needed.
- Be very clear about what a facilitator can, cannot and should not do, and where, when and why to use facilitation.
- Get the basics in place in advance: Timing, venue, refreshments and setting up. None of these should get in the way of the event because of poor planning.
- Focus on the key skills of listening, questioning and reframing, always with an eye to drawing out the positives and establishing common ground.
- Use the support techniques carefully, thoroughly and consistently.

- Aim to predict and prevent problems rather than just letting them happen.

Notes

1. Or, from the thesaurus, 'simplify', 'enable' or 'assist'.
2. To which one could add ethnicity, literacy level and others.
3. Some writers talk of 'wicked problems' as if they have a life of their own.
4. This draws in particular from material produced by Andrew Acland, but as yet unpublished.
5. This draws on material produced by Rowena Harris and Allen Hickling (again unpublished).
6. Once again, this draws in particular from unpublished material produced by Andrew Acland.
7. For more contentious issues and for many familiar with such workshops the latter, in the UK, are often just termed 'Chatham House Rules'.
8. The name appears to come from the way in which rhubarb roots are kept in sacks over winter, or even longer, only to be brought out into the open again sometime later.
9. Some people prefer the term 'holding tray' or even 'parking zone'.
10. This quote is about a project in Totnes in 2012/13. Despite the comment, the process achieved such a high level of agreement that the planning application went through without a single serious objection!
11. This is why this book does not include any suggestion of using traditional public meetings as a method of choice.

References

Australia, Department of Justice (1997) *How to Get the Most out of Planning: A guide to facilitating meetings*. Victoria, Australia: Department of Justice.

Kaner, S. (2014) *The Facilitator's Guide to Participatory Decision Making*. San Francisco: Jossey-Bass.

Mann, A. (2014) *Facilitation – A Manual of Models, Tools and Techniques for Effective Group Working*. Bradford: Resource Productions.

7
Evaluating and Reporting

Introduction

This chapter covers two closely related aspects. The first is forms of evaluation; how processes can be assessed and validated, not just by a client but potentially by all involved. Although the results of this need to be used to inform any final process report, they are also valuable to help to inform future practice as a whole. (This contributes to the key point that evaluation and reporting could well have been presented in reverse order because they are so closely linked.)

The second part of the chapter is about how to prepare final reports of overall processes that can be used, along with technical and content information, to enable some judgement to be made about a plan, strategy or project by an inspector, a committee, a board, a community or even a judge if the legality of a process is challenged.

There is a considerable literature on evaluation generally but comparatively little about evaluating engagement. However, two publications are directly relevant, both by Diane Warburton (Warburton, 2001 and undated). Reporting is simply not covered at all in the literature so no references can be offered.

Evaluation and learning for the future

Putting in place a coherent and well-delivered process, right down to the last detail, is of course crucial. But how does one establish that a process has been well-delivered? For those involved, and then for anybody who wishes or needs to assess the merit of what has taken place, forms of evaluation are essential.

Although most of the following is about evaluation at the end of a process, evaluation at key stages throughout can also be of value for participants, especially in difficult situations where they may need reassurance that things are going in the right direction or where trust needs to be built. In the case of the Guernsey Waste Strategy, a quick evaluation was undertaken at the end of each pair of major workshops. This feedback undoubtedly helped the project team to sharpen the ongoing process and helped participants feel more confident that they were being listened to and that progress was being made. Rather obviously, evaluation at the end is also of benefit to participants, although too often they never receive any feedback to see that their contribution has been valued; the evaluation is seen, wrongly, as something solely for internal use.

However, evaluation should not just be retrospective, about learning from something *in the past*. It is also extremely important for personal and organisational learning and development *for the future*, enabling all involved (which includes participants) to advance their skills on the basis of real evidence and then adapt and improve practice from there on. In addition, although any single evaluation can be used to provide learning for the future, regular and consistently planned evaluation across a number and range of projects can be significantly more valuable in helping to shape and improve future practice as a whole.

Process evaluation

NB. This section is about evaluating the *process* of a project, not the *content*.

This first point will take us back to earlier chapters. Imagine the problem if one were to get to the end of a project, attempt to evaluate it but then had no idea what to evaluate it against, i.e. what criteria or standards to use. Evaluation at any stage is only possible if some principles, standards or objectives are established at the outset. Those principles or objectives may be just those of the engagement managers, they may be those of the clients/commissioners or they can be, as this book recommends, those that emerge from a collaborative exercise with key stakeholders in designing the engagement process in the first place.

Having set something against which success can be checked, the next step is to consider *when* any evaluation should take place, *who leads* that, *what it should cover* and *who contributes* to it.

When?

On small, relatively non-contentious and probably quite quick projects, evaluation only at the end can be appropriate. At the same time, because few processes involve just one event or activity, low-key evaluations of selected events can still be of value. On larger, complex, contentious and hence usually longer processes, stage-by-stage evaluation is of great value (as in the Guernsey example), even if it is still low-key on each occasion. Some form of evaluation at the end ought to be the case for every process, short or long.

Who leads?

Who leads or undertakes evaluation takes us back to the matter of independence. This may not be an issue on small projects or if there is some form of Steering Group who agree when and how to evaluate; the engagement manager just becomes the deliverer of it. For contentious projects, it can be of real value for an external person or group to design and manage the final, if not any interim, evaluation. The evaluation for the Guernsey project was supported by very valuable feedback from an entirely independent group who observed all the workshops and fed back thoughts at all stages (including on completion) to complement those of the participants.

Once again, independence is not an absolute; there may be occasions where someone is brought in by the engagement manager to do it or key parties (or Steering Group) agree who the evaluator might be. The role of external evaluator can be relatively simple, so almost anyone might do it. However, in more challenging situations, any external person will need a good level of skill and experience as an evaluator and, if the work is part of a statutory process, some knowledge of the necessary legal criteria is important. Finally, and to close the circle, it may be possible for members of a Steering Group to act as the overall evaluators, especially if they are semi-outsiders rather than active participants.

What should evaluation cover?

The focus should be on evaluation of the process. Interestingly, experience has shown that people can remain unhappy about the *content* outcome for a plan or project but still evaluate the engagement *process* positively. However, this is difficult to achieve in processes that do not include some form of in-depth work because it is only through deeper engagement that people can fully appreciate all the issues and dynamics at play.

When clear principles or criteria are established at the start, the aspects to address will flow directly from this. There is a need to check, for example, whether people felt the process was managed with appropriate independence, whether it succeeded in engaging all the voices that needed to be heard and whether a suitable range of methods was used. It is nevertheless important not to limit any evaluation to these factors; an opportunity needs to be offered for people to contribute their own extra, different or more detailed criteria.

Who contributes?

The obvious and not entirely wrong (or right) answer is, of course, the participants. However, the list of participants can be extremely long and it may not be possible or even appropriate to try to contact everyone. By including some sort of on-the-spot evaluation within any activity or method, for example, at a workshop, securing feedback from all or most participants is nevertheless possible. This is relatively easy for depth events (a workshop or drop-in) but more difficult for breadth events (a questionnaire) because those who just dip in and out briefly may not be interested in commenting on the process or feel they have time to do so.

Having stressed the importance of evaluating against principles or standards, it is important to be aware that no participant is ever likely to experience, and hence be able to comment on, all aspects or events in an overall process. A different approach is needed when choosing who to involve in an evaluation of an overall process as compared to an evaluation of a specific event.

Considered evaluation is best undertaken with those more fully involved, typically the stakeholders who might have attended workshops. Not only does the whole nature of their involvement create space in which time can be spent on evaluation, but they are also more likely to be able to separate process from content.

The first qualification to the focus on participants, one too often missed, is that in any genuinely collaborative process all parties are (in general) equal, so it is important to also seek feedback from the clients/commissioners, the project technical team and even, if evaluation is being done by others, the process designer and facilitators. It is, however, important to record their judgements separately in case there is any real difference of opinion about success or failure.

The second qualification is challenging but important. There are always some groups or sectors that one has contact details for but, despite all efforts, do not engage. Gaining views from such people or groups, notably about why they did not engage, can be difficult but if you succeed the results can be extremely informative, especially for any next project where their engagement may be more important.

Finally, there is another possible answer to the question who contributes to evaluation? It might be nobody! If a whole process is recorded in a full, final report (see later) and that report includes stage evaluations, it can be appropriate for an external evaluator, especially if skilled, to do what might be called a 'desktop' evaluation, i.e. commenting on what has taken place on the basis of all the material in the report. Though clearly not as good as any evaluation that asks participants directly, this sort of quick testing can be all that some processes need, especially if small and relatively uncontentious.

What methods?

There are almost as many methods of evaluation as methods of engagement. They vary from highly informal, quick and almost casual, to extremely formal, rigorous and time-consuming. They also vary from open to confidential and from one-off to sequential. If there is a high level of conflict at the start of a process, undertaking some form of shared, public evaluation can be inappropriate because how people score is open for all to see. However, as in Guernsey, this can be constructive by showing a high level of openness on behalf of those running the process. By contrast, confidential evaluations, at the end in particular (as also used in Guernsey), can enable people to make comments that they might not feel happy to make openly; yet this way they do not get to see or perhaps learn from the comments others are making.

Judgement is needed about the appropriate style, depth and time needed and this should be proportionate to the scale and complexity of the project and process. Asking people to complete a 20-minute questionnaire form at the end of a two-hour workshop is probably too demanding, but just ticking two boxes at the end of a long series of workshops is almost certainly not enough.

It is also important to add that some organisations or clients actually *require* an evaluation to be done. Most of these organisations are also likely to specify anything from the general manner in which the evaluation is to be done to providing the actual form and a note of who exactly to use it with. Unfortunately, from long experience, these details are almost always inappropriate for process evaluation, so one should be ready to challenge them or perhaps try to ensure that more appropriate questions are added.

As with explaining engagement methods, examples are most informative, as follows.

Workshop(s)

According to the numbers involved and time available, some options for evaluating and reporting on workshops are as follows.

With a small number (12-15), a short discussion can be held at the end. This is effectively a dialogue method, enabling people to make comments, share these with others, hear responses (perhaps very different ones) and hopefully reach agreed conclusions. Some pre-planned and well-structured questions are needed and it can be valuable for someone other than the facilitator to lead the discussion (in which case the facilitator might not, or maybe should not, take part or even be present). The flip side is that open discussion makes it difficult to draw out each person's comments, for example, to compare lay people's views with those of any professionals (although a good facilitator could help with this). Careful notes should be taken and shared back with those present. It is also essential to check whether or not people wish their names to be attached to any report.

With larger numbers (or with a small group) there are two main and not exclusive choices. First, one can ask people for summary reactions on the way out in a semi-public format. This usually involves wall-sheets with scales on which people mark their judgements with a tick (as used at each Guernsey event). They can also make comments on a separate sheet. An example of questions and scales is shown in Figure 26.

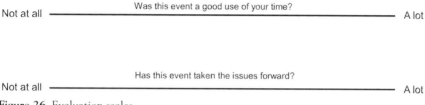

Figure 26 Evaluation scales

The method shown in Figure 26 is quick and easy and enables everybody to see the results before they leave (although the results should also be in the event report). The method has the advantage, though also the disadvantage, of not being able to attribute responses to any individual or group.

The second choice is to give all participants an evaluation form. This can be confidential to encourage critical comment and, if so, people need to be told that. In the same way, if the results are to be shared with all participants or passed on to others, that too needs to be made very clear. Using postcodes or perhaps organisation name can make it semi-confidential, and helps to show any trends in who evaluated in what way, for example, pedestrians being unhappier than cyclists. For any single event the form should be relatively short, requiring no more than two or three minutes to complete. A more thorough and lengthy format might be used at the end of a series of workshops, in which case it is important to warn people of this and allow more time in the final event programme. The following is a slightly edited version of the form used at the end of the final waste strategy workshop in Guernsey (the original included more space than shown here for comments on each question):

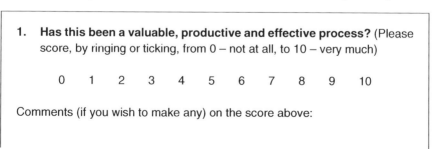

2. **Do you believe that the consultation process will inform the development of the eventual waste strategy?** (Please score, by ringing or ticking, from 0 – not at all, to 10 – very much)

      0   1   2   3   4   5   6   7   8   9   10

Comments (if you wish to make any) on the score above:

3. **Do you consider that your voice and the voice of those on whose behalf you have been present have been properly heard?** (Please score, by ringing or ticking, from 0 – not at all, to 10 – very much)

      0   1   2   3   4   5   6   7   8   9   10

Comments (if you wish to make any) on the score above:

4. **Have the workshops been appropriately engaging, informative and active?** (Please score, by ringing or ticking, from 0 – not at all, to 10 – very much)

      0   1   2   3   4   5   6   7   8   9   10

Comments (if you wish to make any) on the score above:

And ... any other comments or points you wish to make:

As suggested, the two approaches are not exclusive; one could do a very quick ticking exercise to get an overall and shared assessment, after which people could reflect on that as they complete their forms.

Drop-ins etc.

By and large, people attend and contribute to these as individuals, even if they ostensibly attend on behalf of an organisation or come with friends. One can use a version of the simple scales and comments sheet as in the example used in Guernsey. This can be successful because people often like to see what others think or have commented on. There can, however, be practical issues if very large numbers are involved and, by definition, whoever dashes off first never gets to see the final result (until it is in the final report).

If using a response form as a way for people to make comments on the content of anything being displayed, you can use the same form to present questions about the process. If at all possible, such forms should be completed on the spot; beware of those who say *'I'll take it away and get it back to you'* because they almost never do!

An example of a question included on a response form follows below. This was created for the final public event in a series used (with several other methods) for engagement on a housing project in Totnes. The event was part display, part final consultation, fully open to the public. What follows is only the process question:

Response to consultation comments

- The key points from consultation with local people are listed on each display board along with the design team responses.
- Please comment here on how well you think the designers have (or have not) picked up on what people in general (not necessarily yourself) have been saying.
- Tick/ring a term below and add comments if you wish.

Not picked up well	Picked up only a little	Some picked up, some not	Picked up quite a lot	Picked up very well

Comments:

In this case respondents were asked just for their postcode, enabling some valuable analysis of comparative responses from, for example, those living very close to the site as compared to those living further away.

Questionnaires etc.

For consultation that involves anything at distance, notably questionnaires, Twitter and other electronic or online methods, there is no real choice other than using another questionnaire. This could involve asking participants to complete a confidential feedback survey via a project website or asking a sample of participants (provided one has contact details) to fill in a hardcopy and/or electronic/online questionnaire. Response rates are often even lower for evaluation questionnaires than with questionnaires used to address the content of a process. In addition, as they are closed and confidential, one can get some more forceful (even abusive) comments than with other methods. The content of the questions to ask is probably similar to that in the example shown for a drop-in.

Feedback

Providing feedback at all stages was introduced as a key principle in Chapter 2. It applies just as much to giving feedback of the results of any process evaluation as it does to feedback about content results from a workshop or questionnaire. If the feedback is to be to each respondent, that is dependent on having their contact details, which one may not have even requested, so plan ahead if you wish to do this. For anything done via websites, social media sites (Twitter etc.), personalised responses are not possible but the methods themselves enable results to be posted alongside everything else. Even if personalised feedback is to be used, it is also always valuable to place the results of any evaluation in the public domain, for example, on a project website.

Confidentiality is a very important issue here. It is not common practice to list names of contributors to evaluations but, if this is seen to be valuable, it can only be done with full approval of those involved. Comments supplied often provide very valuable potential quotations. These can be confidential but again, if it is thought valuable to name someone, that needs to be checked. Finally, of course, the evaluations should all form part of a final process report as described in the second part of this chapter.

Reporting: Getting your work valued

Having addressed everything raised in previous chapters, the last thing one would wish to see happen is for somebody in a decision-making role to be unaware of, ignore or simply not support the work undertaken and its outcomes. In many situations there is some person or group (client, Committee, Board, external Inspector) with a formal or informal role to assess or examine the plan, strategy or project around which engagement work has focused. It is therefore essential that they receive a final report of the engagement activity.

In fact, this is becoming more necessary as consultation or involvement becomes more formally embedded in English legislation and guidance as an integral part of developing plans or projects (and in some areas of policy other than planning). This greater status for engagement then brings with it a related requirement to present evidence in a way that enables any assessor to give it the most appropriate weight. Taking this further, there can also be a need to ensure that any engagement process and its conclusions can stand up to legal challenge.[1]

As yet there is a total absence of guidance in the planning world on how to prepare final reports in a way that maximises the chances that their contents will be properly valued. This is worrying given the already existing requirement for such reports to be submitted alongside all other information for examination of any statutory plan. Perhaps the nearest one can get to guidance comes from the Localism Act (Great Britain, 2011). In relation to what is required to be submitted with any Neighbourhood Development Plan, the Act outlines the following about what is termed a 'Consultation Statement':[2]

(2) In this regulation "consultation statement" means a document which—
   (a) contains details of the persons and bodies who were consulted about the proposed neighbourhood development plan;
   (b) explains how they were consulted;
   (c) summarises the main issues and concerns raised by the consultees; and
   (d) describes how these issues and concerns have been addressed in the proposal.

This is sparse but valuable and covers, as most other sections of planning regulations do not, the more informal stages of consultation as covered in this book. However, similar regulations in the past have not been followed up with more specific guidance and the current (2014) government is generally disinclined to provide guidance.

In addition, even with the best possible report, the value given to it depends very much on the ability of the person assessing it and on the weight they choose to give, or feel they are required by legislation to give, to any engagement outcomes. Here too there is a worrying gap because the vast majority of planning officers, Inspectors, councillors and others have no training at all in how to assess the quality of any engagement process. Worse still, the anecdotal evidence is that, despite the apparently rising status of engagement, little or no weight is given to engagement results when decisions are made, and practice appears to vary considerably between, for example, Inspectors. Change is slow and it can be argued that one way to speed up the change is through the preparation and submission of good and thorough reports. Good reports can illustrate the parameters of high quality engagement (thus helping those as yet untrained) and they can present strong evidence to which real weight can be given.

Given the absence of any useful guidance in the world of planning, one can turn to more general legal processes to find the precedents and hence the standards that the world of consultation practice uses and which are supported during a legal challenge. Although the principles and standards in this book are consistent with these standards, and what is in any report should relate to them and show how they have been met, those standards as yet apply only to the *formal* stages of consultation. This is all covered more fully in Appendix 2.

As for evaluations, there is then the question of who writes and/or who validates any final report. If a process is transparently designed and delivered with an appropriate degree of independence, the process manager can draft it. A report should be clear enough and comprehensive enough for others to make judgements based on it, even judgements that differ from those of the report's author. If there are likely to be questions about the independence of the report's author, or if significant disagreements are still unresolved, two other approaches are possible. First, some form of Steering Group can be established to either test and sign off the process manager's report (as in Totnes) or produce their own (as in Guernsey). Second, the process itself, or the manager's report on it, can be evaluated separately by someone appointed by client, community, local authority or other who is acting completely independently (as this author has been commissioned to do on occasion).

Finally, and very obviously, it is impossible to produce a robust, defensible report from a poorly planned and poorly delivered process! At best, a good final report merely captures and builds upon all the good decisions taken in designing and delivering a genuinely collaborative process. In other words, all of the questions one asks at the outset, or in designing any specific event, need to be reproduced, with their answers and the process outcomes, in a final report.

Shaping a good report

To provide a flavour of what needs to be in a robust final report, imagine being an independent evaluator of an engagement process and reading this:

'*A meeting with stakeholders was held on 27th April to discuss the final proposals. The proposals were agreed.*'

This raises some obvious questions:

- Whose meeting was it?
- Who were the stakeholders?
- Was the 27th a weekday or a weekend day? And which year?
- Where was the meeting held?
- What proposals were presented?
- What format was used to assess stakeholder responses?
- What does 'agreed' actually mean?

By contrast, look at the following first section of the report of the first workshop for the Waterton project:

> This workshop was held at the ... Centre on 22nd April 20—. It was the very first event in what will become a wide-ranging programme of consultation/engagement activities to support the development of the masterplan for the Waterton project. The workshop was targeted (by invitation only) at a group of around 60 people who, either as individuals or through their group or organisation, were seen as key 'stakeholders' in the project. This group, which will meet at least once more, is now called the 'Forum'.
>
> The list of invitations was put together by ... (who is managing the consultation work) with assistance from officers in ... District Council. Everyone on the list was invited to attend the workshop. Those booking-in received, with their joining instructions, an initial briefing note about the project to help them understand the context.
>
> The final participants list is included at the end of this report and includes those who sent apologies. On reflection, given the relative novelty of this project, it would have been valuable to send the briefing note with the initial invitation in order to help people understand the project better. This might have encouraged more to attend, although the number on the evening (42) was excellent both in total and in diversity.
>
> Following a short introduction by the facilitator (also mentioning project team members present – see later list) the workshop went through two main stages:
>
> - Issues Sheets and responses
> - Group 'Plans'.
>
> Each of the two stages is introduced fully later in this report, and all results are included. All text formatted in italics – as here – is introduction and explanation, everything in plain text is as used or recorded on the evening.
>
> The main points from the report will be shared amongst the project team, the client group, the Steering Group and the Partnership Board. They will be used (in fact, they are already being used) to inform the next stages of consultation. Those next stages were mentioned briefly by the facilitator at the end of the workshop but have been repeated here in the final section in case details were missed.

No report can ever comment on every detail that an evaluator (or a barrister) might query but the Waterton example provides answers to most of the key points and demonstrates that things were done properly at the time, not picked up retrospectively.

A brief halt is necessary here for two key reasons. The first is to stress the point that a report is an *integral* part of any process. As per one of the principles outlined earlier, the process to be used would ideally be agreed at the outset with potential key players. It is difficult to over-emphasise the importance of this. If the process is agreed in advance that is a stake in the ground because it ensures that the process itself is less likely to be challenged later. If the process is delivered broadly as agreed then that is a second stake in the ground; the basic delivery can hardly be challenged. That leaves the detail of the delivery, certainly challengeable, and the use made of results, also challengeable. Having two out of the four agreed is a huge leap forward and those in the planning world will realise that this is exactly what happens, for exactly the same reasons, in the scoping stage of Environmental Impact Assessment. Very simply, do not leave thinking about the final report until the end of a process; build it in from the start.

Second, and very practically, lengthy and coherent engagement processes will generate a great deal of material. From this author's bitter personal experience it can be traumatic to wait until the end to start ploughing back through everything to build a report. Instead, start early and keep logging output material, session reports, press releases or whatever as work continues, maybe not weekly but certainly at key stages. That regular piecemeal reporting can also be extremely informative about things that may have been missed or about things that may need to be changed.

Building blocks of a final report

As with many documents in and around planning and other policy areas, there is potentially a considerable amount to include in a final report and as yet there is no standard way of doing one. What follows is based on this author's proven experience, including reports on processes that have successfully survived legal challenge. For reports on technical issues, for example, Environmental Impact Assessment, the usual format is a summary report, if a quite thorough one, and full Technical Appendices. That is the basic approach recommended here for a final 'Statement of Community Involvement'.

A **summary report** should include the following:

- An introduction to explain the purpose, role, structure, contents and authorship of the report and the links between the Summary and the Appendices.
- A full repeat of, or key extracts from, any requirements and/or guidance on engagement.
- A brief outline of the plan or project, focusing on key aspects addressed by the engagement.
- A note on any previous consultation work.

- A note on how the programme was prepared, especially if that was widely agreed.
- An outline of the overall programme of activities as planned at the outset.
- Comments on what actually took place given that this may have changed from what was planned, and notes on why and in what ways things changed.
- A section on each main activity, summarising how it was managed and key results.
- A note on the key issues that emerged, followed by a description of how the plan/project *has* responded to the engagement results and a description of the main items on which the plan/project *has not* responded, with a brief explanation of why it has not done so.

These are the basics. If the report is to be self-evaluated, there is value in including a final, reflective and strictly personal evaluation by the engagement manager of the strengths and weaknesses of the approach, for example, key groups not adequately accessed, events poorly attended, particularly successful activities or outstanding issues for any next stages. If this is done, it is essential to make it clear that this section is distinct and different. Honest comments at this stage can be (and have been shown to be) an effective way of limiting or reducing the impact of later challenges and can also have real value in helping an assessor (especially if untrained in engagement) to reach conclusions about the quality of the process and its outcomes.

What follows is a series of key points, sometimes with examples, about the content of the elements suggested thus far. The Appendix or Appendices referred to in the Summary would simply be a file/folder containing all reports (e.g. of workshops), notes (e.g. of meetings) or examples (e.g. of website pages).

Introduction

An example is useful here, based on a report about a housing project in Modbury, Devon:

> 1.1 As part of the work undertaken on the proposals for the RA1 site in Modbury, there was a planned programme of community engagement, consultation and communications. In order to explain and appraise the proposals, a thorough record – an audit trail – has been kept from the start of the commissioned process. This report and the accompanying Appendix describe that audit trail by covering what took place, the background to it and the summary results. The final section includes an audit of the application proposals against the key results from the engagement and consultation work.

> 1.2 The overall report is in two parts. This **summary report** sets the scene and describes the process in broad terms, focusing on the overall pattern of work, the main activities, key results and the flow from initial issues to the final proposals. It is backed up by the second part, an **Appendix** that contains specific and full reports from workshops, actual results from exhibitions, basic information about publicity, ad hoc responses and so forth. The latter includes all the verbatim, raw material to which this report is the summary.
>
> 1.3 This report was put together by ...
>
> 1.4 Where links are made to items in the Appendix they are shown as follows: **ITEM X**

Requirements and guidance

This should cover national policy (by reference to an Act), local policy (by reference to the relevant authority's Statement of Community Involvement), less formal local policy or guidance and perhaps any other important material. In one case the latter was a recently published national report by a wide range of groups promoting better engagement.

The plan or project

This need only be short; simply a description of the nature of the project, clients, timescale, key issues, previous history etc., i.e. anything that might be seen to then inform the engagement programme.

Previous consultation

As explained in Chapter 3, there will often have been previous consultation processes, perhaps directly (on the specific plan/project), perhaps indirectly (on a project within a plan that was consulted upon) or perhaps just in the general area (but still seen by some as relevant). There is little need to describe these in detail but reference should be made to them, to any lessons learned and to the way in which they might have set the scene for what follows.

Preparation of a programme

This needs to describe how the programme was developed. It can be very short if the task was done solely by the consultation manager. If it was presented to and only generally discussed with others, such as the client or the local authority, that needs to be mentioned. Most importantly, if the process was developed and agreed with others such as key stakeholders, this procedure needs to be described, perhaps supported in the Appendix with a report of a process design workshop. If that initial work resulted in the formation of an ongoing Steering Group, that should be mentioned here and then picked as a section in the main part of the summary report that lists the main activities. Ideally this should include notes of Steering Group meetings.

The programme as planned and as delivered

Because each activity will be described in the main section, this need only be a summarised list of what was planned at the outset and, depending on the degree of variation from the original plan, notes of any changes. If things did change – timetable, project content, new groups forming, workshop dropped, meeting added etc. – brief explanations need to be given.

The main section: The activities

Each of the activities in or elements of the engagement programme needs to be addressed. In the main, this is best done in chronological order although some activities, for example, a website, may have continued throughout. The latter are best left to the end in order to report on final results.

The section on each activity typically needs to include the timing of the activity, its format, who was involved, general success (or otherwise) and key results. Each section should refer clearly to the full reports in the Appendix. Aspects to be covered for typical activities are outlined here (and are broadly similar to those in the example in the box on page 163).

For a **workshop** it is important to cover:

- The general context (stage in the programme).
- Aims and outcomes.
- Invitees.
- Date, location, and time.

- Who actually attended (including any of the project team).
- The overall format/programme.
- Key results (usually copied straight from the full event report).

For an **exhibition** or **drop-in,** the list is similar, substituting a note on how the event was advertised for details of invitees, noting total numbers attending and, if possible, how long people stayed on average.

Questionnaires are most often a specialist and therefore sub-contracted task, so the task managers would have produced the report. In summarising, it is important to cover the following:

- General context, aims and outcomes (as above).
- Sample basis, selection criteria and methods to be used.
- Dates and times for completion.
- Methods used.
- Sample achieved or not achieved (in which case why not and what if anything was done to address this).
- Key results (as above).

There is no obvious or simple way to summarise activities such as **awareness-raising** and **publicity**. It is best to describe each activity, picking up on when and how it was used and with or for whom, and then illustrate this in the Appendix with a few examples: A press release, advert, newspaper article etc. If any of this generates results, for example, letters to a newspaper, emails to the engagement managers or a Twitter exchange, key points can be drawn from these and the full notes included in the Appendix.

Websites are often used throughout a process. If they are passive, i.e. not offering any opportunity for input, the summary would need to highlight when the site was set up (and closed down) and what in general it covered at what stage (for example, offering downloads of event reports). If the site is interactive, the format for this should be outlined and some summary points from feedback included. It is valuable to include some or all main website pages in the Appendix.

Response to consultation results

Reporting on the aspects already covered can be relatively objective and solely about input, i.e. what consultees and stakeholders have offered as ideas, issues, comments or queries. None of it addresses how, if at all, and to what extent the plan or project has responded to this input. This is where the balance of any report starts to change. It is of course possible to stop at this point and leave it to whoever receives the report to compare the results to the final plan or project. If done with care, however, it can be valuable to include a section that highlights obviously positive responses and perhaps some queries; readers of the report can still make their own judgements.

In one case (Modbury) this section was introduced with the following statement:

> All we can do here is to highlight simply that points have been addressed in some way; whether or not they have been addressed fully or satisfactorily is for others to decide during the application process. We also comment on some potentially important issues that were raised and to which the project has perhaps not responded fully, offering explanations provided by the design team (not by ourselves).

One example of a format for this follows here, from a small housing project in Salcombe, Devon:

Community key point	Design response
1. New development should be sensitive to the rural setting through tree planting and careful layout taking into account topography.	• Proposed layout of buildings and landscape follow contours. • Use of predominantly native tree planting to visually relate to the adjoining landscape.
2. The development should respect and reinforce the rural edge to ... , the wood between the site and ... Road, together with the fields between ... and ... which are important 'gaps' and edges to both distinct settlements.	• Importance of the surrounding woodland to the character of the new development and the wider character of the area is acknowledged and drawn on in the proposal. • Retention and enhancement of existing boundary hedgerows with a suitable landscape strategy for their future management.
3. New planting will be an important way to soften views to the development, retaining existing woodland and planting new trees across the site.	• Creation of a strong landscape structure for new development, with bands of tree planting along access roads and trees within parking areas, to soften views from the south and visually integrate proposed development with the existing wooded setting of ... , as well as providing some shelter.
4. The development should be linked to the town with safe pedestrian routes.	• Footpath provided to the east, adjacent to the cemetery, to enable people to choose a variety of pedestrian routes into town. • Connection to cycleway along the A ... leading via ... Road to the school and to the town centre.

Community key point	Design response
5. Integration into the existing landscape is important, with careful landscape design (new and existing vegetation) and siting of buildings to reduce visual impact.	• Use of predominantly native tree planting to visually relate to the adjoining landscape. • The general housing layout closely follows the contours.
6. Access to the development from the existing entrances accepted.	• Access to the main site uses the existing entrance point.
7. Provide 'affordable' housing that meets local needs, both in terms of tenure (rental accommodation needed) and house types (e.g. 2–3 bed).	• … Homes have sought and used advice on this from … They identified a requirement for one and two bedroom properties to be provided as affordable housing for local people who are otherwise unable to buy or rent at market values.

Process evaluation

In addition, having focused on project content, any evaluation done of the process can be mentioned at this point and summary conclusions added in. Drawing on the simple scale method used earlier in this chapter (page 159), the results could be summed up as follows:

Satisfaction with the process

The final part of the Response Form used at Drop-in 2 asked people to comment on how well they felt the team had picked up on their input. They were asked to tick or ring whichever of the following statements applied best in their view:

        1 – Not picked up well   2 – Picked up only a little

3 – Some picked up,   4 – Picked up quite a lot   5 – Picked up
    some not   very well

40 people did this, scoring as follows:

      1 = 4 people   2 = 2   3 = 7   4 = 19   5 = 8

This averages out at a score of *c.*3.6.

Final reflection

If one wished to do more and include a more personal reflection by the process management team, it can be valuable to go back to the original principles or standards used in order to comment on how the work undertaken relates to these. As stated earlier, this must be done carefully and cautiously and its authorship made very clear. A selection only of text from the final report on the Waterton project follows to illustrate how this might be done:

> The District Council's Statement of Community Involvement includes the following aims or principles of involvement. These are listed in bold below, followed by the engagement team's own thoughts on how the process relates to these.
>
> (Selected examples only)
>
> - In terms of **Informing**, we have commented on occasion about the problems of establishing a continuous flow of general public information. This is the bedrock on which project-specific involvement can be built; without it one is always struggling to even begin a process, as happened here (see also next point).
> - In terms of **Inviting**, this report shows that a considerable amount of effort was put into raising awareness and generating interest and that different people were offered, in several ways, several opportunities to be involved. The low level of subsequent responses may result from the nature and quality of what went out, or the ways it reached (or did not reach) people. This may be part of the explanation but we believe it is not all of it. One must look elsewhere for other explanations, perhaps back to the point in … above about a lack of 'bedrock'. We must, however, point out that any lack of response should not affect how the results from the 1,200 or so who did engage are judged.
> - In terms of **Involving** people, the events that were run all provided opportunities, at what we consider to be the appropriate (front-loaded) stages, for people to raise issues, offer information and help (to varying degrees) to develop options and final proposals.

Key messages

- Evaluation of any process is important, for clients, participants and for those managing the process, and for any final report and those who receive it.
- To do this successfully requires clear principles, criteria etc. to have been established at the outset.

- The level of detail, methods chosen and who undertakes any evaluation must be appropriate to the scale and contentiousness of the project.
- Stage/event evaluations can be important; final evaluation is essential.
- Think widely and carefully about who contributes to any evaluation.
- There are many methods of evaluation; choose and adapt methods as appropriate to the specific context.
- Be sure to feed back any results of evaluations.
- Final reports are valuable anyway but are becoming more valuable, more important and more often challenged as engagement becomes more common, and especially if it is or becomes a legal requirement.
- Final reports should be appropriate to their project context.
- A proven format involves a summary report and some form of Appendices, the latter containing all the full event/activity reports, i.e. a complete audit trail.
- It is good to think about what needs to be included that would answer any critic's queries, especially those of a barrister.
- Overall evaluations of the impact of the engagement on the plan or project and on the process are needed but should be clearly stated to be more subjective.

Notes

1 See Appendix 2 for possible legal challenge criteria.
2 This is part of the Neighbourhood Planning (General) Regulations. Go to: http://www.legislation.gov.uk/uksi/2012/637/part/5/made

References

Great Britain, Department of Communities and Local Government (2011) *Localism Act*. London.

Warburton, D. (2001) *Evaluating Participatory, Deliberative and Co-operative Ways of Working*. InterAct Working Paper. Unpublished but contact this book's author at: jeff@placestudio.com

Warburton, D. (undated) *Making a Difference: A guide to evaluating public participation in central government*. London: Department of Constitutional Affairs.

8
Making it Mainstream

Introduction

So far, this book has focused on how to design and deliver successful processes around single issues such as a strategy, plan or project. This is appropriate but in some ways not good enough. It could in fact be damaging, even wasteful, to create stakeholder lists and processes every time a new challenge arises.

There are two key factors behind this point. First, for many commissioners of engagement – local authorities, public agencies, developers and others – there are likely to be several consultative exercises under way at any time. Treating each as completely separate can waste time and resources and potentially confuse, even put off, potential participants. Second, treating every situation as a one-off misses out on the potential benefits of mutual learning, practice and experience for commissioners and participants over time and across many projects, i.e. what is often called capacity building. Given a coherent and consistent overall approach, rather than starting off every new task at the very beginning again, one starts a step or two up the ladder each time.

This author's experience offers a worrying example of what can happen if some of this is not done:

> In one local authority, around 25 people from that authority, local agencies and communities came together to list all the engagement activities that had taken place in their area in the last six months, those under way at the time and those they knew were planned for the next six months. They listed 128 engagement initiatives but …

- Almost all shared some content, none was completely discrete.
- The principles and methods used had no common basis.
- Some stakeholders were invitees on all 128 initiatives.
- On several occasions, three or four events had happened on the same evening.
- Nobody had any idea how much each event, or any total process, had cost the authority or the participants.

- There was no process, no person responsible for managing all this; it was 128 one-offs!

This session led to a discussion about what is termed 'consultation fatigue' amongst both participants and providers, which everyone agreed to be a perfectly sensible and understandable response to this fragmented and unplanned approach. Compare this to another of this author's experiences:

- In one District (District A), a coherent programme of engagement was put in place.
- This had an implicit capacity-building dimension through which many of the key officers, elected members and stakeholder representatives in that authority slowly developed their ability to understand and contribute to engagement work.
- Engagement then started on a Waste Strategy for the whole county, covering not just District A but six others.
- The initial workshops with stakeholders in six of the Districts went fairly well and then it was time for the session in District A.
- Because everyone involved from District A was familiar with engagement principles and methods, and with working productively with each other, the county team agreed that more and more valuable results emerged from that one event than from the other six put together!

Most readers will, however, have limited opportunities to implement the ideas covered in this chapter because few are likely to be in a position to do anything about building more coherent, organisation-wide approaches to engagement practice and shaping a genuinely coherent programme. Yet this is patently needed if high quality practice is to become the norm rather than the exception, i.e. to *make it mainstream.*

This chapter addresses two aspects of this challenge. The first is fundamental. There is little chance of achieving consistent, appropriate and cumulative delivery of good engagement if the organisation responsible for it is not culturally committed to it, if those involved do not have the appropriate skills and if the appropriate resources (in all senses) are unavailable. Some level of 'capacity' on engagement issues across a whole organisation is essential. The second level is about the framework within which to maximise any raised capacity. This requires, amongst other things, a strategy, a set of principles and a managed programme. None of this is easy and it is all as yet poorly covered at any level from theory to practical guidance to proven practice.

In addition, moving these ideas forward is not a task for one person or even a few, especially if they are, as is too often the case, relatively junior. 'Making it mainstream' is the responsibility of organisations, be they local government or development companies, even large NGOs.[1] As the following section on organisational capacity shows, this book would be reneging on its responsibility if it failed to address this broader context. Without some broader level approach, the value and impact of even the best designed and best delivered single engagement project is limited.

This chapter therefore covers:

- **Organisational capacity building for engagement**: An outline of what it would mean for an organisation to have a coherent approach.
- **Building capacity**: Some processes to put in place that can help to build capacity for engagement.
- **Procedures and structures**: Assuming capacity and commitment, this section outlines what needs to be in place to deliver on that expanded capacity.

Using the cooking analogy for one last time, one could say that this chapter is about the equivalent of ensuring an overall, regionally distinctive 'cuisine' (as we are told is the case in Italy and France).

Organisational capacity building for engagement

Bristol's positive example has already been looked at in Chapter 2, i.e. the engagement for the regeneration of a park, and we will look at another positive example in Bristol later. However, while the park engagement process was under way, simultaneously consultation was started in the same area by another team on a related set of issues, but in complete isolation from the overall park project. Why is it that, even in the best managed settings, such unfortunate examples can still happen?

There is certainly the inevitable human error that seems always to accompany ambitious ideas, such as consistent and coherent city-wide consultation and collaborative working. There are, however, deeper and more insidious forces at work related to the culture of organisations and professions. As has been pointed out so incisively,[2] the planning system is exactly the same across England, yet there are exciting and positive new developments in some places and appalling examples in other places. The explanation cannot therefore lie with the system but with the different cultures of officers and members (and perhaps even communities and NGOs) in those different places. (This is also not a factor of differential wealth; some of the best examples come from lower income areas.)

What is it about the culture of organisations that ensures success with their planning and environmental processes? And what does or does not happen in the others? Even if one could find an answer, how does one change or shift the culture of the poor-performing organisations? Issues of organisational behaviour and culture change as a whole are far too complex for this book, especially if one overlays the particular, still novel and challenging focus on better engagement practice. Change for a local authority in how it delivers on a familiar topic such as waste management is difficult enough, or for a developer on a topic such as building regulations, but engagement and collaborative working are, for most, totally new, belong to no established profession and group and are as yet impossible to assess in traditional ways such as cost–benefit analysis.

This relates back to the list of barriers to collaborative working outlined towards the end of Chapter 2. Barriers such as professionalism, the challenge of providing evidence of benefits and narrow notions of career development, apply particularly strongly to large organisations (private as well as public), to which one could add departmentalism (or departmental protectionism), hierarchies (rigid levels of decision-making and discretion) and managerialism (valuing internal efficiency above external effectiveness). And again as outlined earlier, some key characters are involved, most importantly the elected representatives.

However, there has been some recent attempts at organisational capacity building for engagement (if not specifically in planning) which at least offer some useful insights, warnings and hints at what to try (or avoid), if not yet any truly convincing results (see Bull, Pett and Evans, 2008, Colbourne, 2010 and Cuthill and Fien, 2005).

Towards a model

What might be the form and nature of a genuinely collaborative organisation that enables more, better and especially more coherent and consistent practice to take place?

The diagram in Figure 27 shows a model developed by this author (Bishop, 2005). It has been used in a number of different organisations to begin a debate at senior level. It applies mainly to local authorities but it could also be applied to national government, government agencies, very local government (Town and Parish Councils) and NGOs, if always with caution and some variations not covered here. It could, probably should, also apply to private companies such as developers. (For any private sector body, substitute Board for Authority members.)

(The focus on local authorities is partly done to simplify the picture. According to true collaborative principles, of course, there ought to be just one overall diagram for everybody. Please bear with this bending of the key principles; it is no more than a means to an end.)

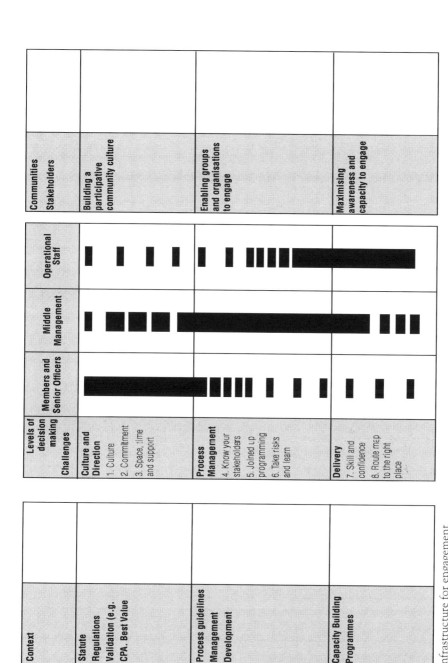

Figure 27 Infrastructure for engagement

The central part of the diagram relates to a single authority. To the left are the input, influence and perhaps even involvement of central government (or national bodies such as the Environment Agency). To the right are the input, influence, and in this case definitely the involvement, of communities and stakeholders. (With a private sector body leading, they would occupy the central part of the diagram along with the decision-making authority.)

The central part of the diagram is divided vertically and horizontally. Vertical divisions relate, in simplified form, to the different levels in the organisation, from the **senior/top** level of elected members/board and chief officers, through **middle management** to the more **junior** staff who actually deliver engagement. Horizontal divisions illustrate the following:

- The top **Culture and Direction Zone** refers to all the ways in which those leading or managing an organisation can illustrate, embed and communicate their organisation's general commitment to high quality engagement. That sets the scene for all others *internally*, giving them the confidence to know that their own small contribution is valued throughout the organisation. It also makes a positive statement *externally* about that commitment to developers, communities, consultants and stakeholders.
- The next level, the **Process Management Zone**, is where the overall commitment is translated into coherent, consistent, programmed, timetabled, budgeted, staffed, monitored and evaluated collaborative processes and projects, i.e. not just a set of random one-offs.
- The third level is termed the **Delivery Zone** because this is where the real work, the actual engagement activity, happens, even if those at other levels are also engaged (which they should be).

The vertical black bars highlight lead responsibilities; a genuine lead role where the bars are solid (e.g. members and senior management with their prime responsibility for culture and direction) and a need to have a presence and gain some direct experience at all levels where the bars are broken. Note that it is not just the coal-face workers in the Delivery Zone who have a responsibility to reflect and convey the overall organisational commitment in everything they do; members and senior management should also reflect that.

The box to the left in the diagram, titled **Context**, relates mainly (in this version) to central government. It is less fully explored here but is included to avoid suggesting that local authorities operate in a vacuum. As elaborated in Chapter 2, most authorities operate within frameworks or regulations, formal or informal guidance passed down to them from central government, from Europe or even internationally. What comes from these levels can also be specific and hence impact on the middle level of the diagram. There are also national programmes under way (for example, the training provided by the

Planning Advisory Service) that impact at the lower levels. What comes down from government and others can be carrot (help/guidance) as well as stick (standards/regulation) and this is subject to change according to the government in power at the time.

The box to the right, **Communities and Stakeholders**, completes the suite by pointing out that action within or by organisations is never enough. Creating and promoting an *internal* culture of participation is valuable but that culture needs to be understood and shared very widely with all others *externally*. This area is again less fully elaborated here and is so large, complex and diverse that it can be dangerous to even give it a single label. Note also that the content of this column is rather different because it mentions the sort of background activity that a local authority can lead on or help with to prepare people for collaborative working, for example, initiatives such as citizenship in schools and community capacity building.

The diagram in operation

The central part of the diagram has 12 boxes.

- Action *only* in the bottom right hand box will have limited overall or cumulative value. It could be called 'adhoccery' because, without broader and higher level support, the best one can do are some good one-offs but without ever adding to organisational capacity. Reading the majority of guides to engagement, this might even be described as the 'techniques blind alley', a dangerous focus on methods/techniques ('ingredients') without an equivalent focus on processes ('recipes' and 'menus'). However, through no fault of their own, many highly motivated and skilful people often become trapped in this particular blind alley.
- Action *only* in the top left box could be called 'gestures bereft of action', the domain of empty commitments, vacuous public statements and policy written but not delivered. Corporate statements to the wider public not backed up by action at the coal-face are a classic example of what gives government its poor reputation. Top level commitment is only as good as what a junior officer says to a worried member of the public at a consultation event.
- Experience from other areas of practice suggests that it is too often at the middle process management level that the real change of attitude is needed. At worst, middle managers block the route from top level commitment down to action and prevent bottom level action from filtering up into general practice. They are classic 'gatekeepers', controlling the crucial policy, plan or project-specific staffing, timing and budgetary decisions and they also have a major influence over training programmes and budgets. Without them genuinely on board, no real progress is possible.

If good progress is only possible if all the boxes are filled, what actually happens in practice? A few years ago a group of 50 or so engagement advocates came together to answer this question. Most were practicing process designers and facilitators, others were community development workers, trainers and staff from government departments and agencies. The results, expressed in summary in the diagram in Figure 28, were clear, consistent and chilling.

The shaded boxes signify where most people judged that there was *some or a lot of positive action* under way. Unshaded boxes signify where almost everybody stated that their experience showed *little or no positive action* in that box. Quite bluntly, this shows that 'adhoccery' rules! With one or two exceptions, nobody was able to offer a strong example of high level commitment and there was disappointment, sometimes anger, at the lack of middle management support. Things have progressed a little since the workshop (as we will see later) but it would be risky to suggest that good practice that enables all boxes to be shaded is still anything other than the rare exception.

Note, however, the shaded boxes to the left under the contextual or government levels. Several people explained this by saying that what was coming from government at the time by way of regulation and guidance may not have been as good as they might have liked, but there was more than enough of it to reasonably expect more and better practice at local authority level. More regulation, better guidance etc. was therefore seen to be valuable but not the answer.

Capacity building

Capacity building is the term used to describe not just the ways in which individuals and groups can develop their awareness, knowledge and skills but also how they can build their self-confidence and their ability to maximise new skills in many different settings, probably with different people. Within an organisational context it also has a key dimension of binding everybody into a shared set of commitments and approaches, i.e. an overall organisational culture.

This comes, however, with an important warning. The phrase 'capacity building' tends to imply that there is little capacity present at the outset. This may be true on some new topics but it can be a dangerous, almost insulting, assumption to make. This is because there are often people within an organisation, and in the community, in companies etc., with considerable capacity, but that capacity is not recognised or fully used, as with the skilled and committed junior staff mentioned earlier. Some commentators prefer the term 'capacity release', implying building from and releasing what is there already and which may be being held back by organisational blocks.

Context		Levels of decision making Challenges	Members and Senior Officers	Middle Management	Operational Staff	Communities Stakeholders	
Statute Regulations Validation (e.g. CPA. Best Value)		**Culture and Direction** 1. Culture 2. Commitment 3. Space, time and support				Building a participative community culture	
Process guidelines Management Development		**Process Management** 4. Know your stakeholders 5. Joined up programming 6. Take risks and learn				Enabling groups and organisations to engage	
Capacity Building Programmes		**Delivery** 7. Skill and confidence 8. Route map to the right place				Maximising awareness and capacity to engage	

Figure 28 The model as experienced

Organisational capacity building is still a relatively young area of work and doing it on a precise, bounded technical issue such as information technology is totally different (and far easier) from doing it on something as wide-ranging and complex as engagement. No entirely convincing and/or proven models exist but blending together what some people have been working on suggests the following general framework.[3]

Work across the whole organisation, not just a few individuals

As in the model presented earlier, any programme of capacity building for engagement should apply to all levels in an organisation. Preparing and managing a programme cannot be left to a senior officer's task group. It must involve people from all areas and levels, if not on a regular, working group basis but certainly in a way that ensures that any programme reaches all parts of the organisation. That should also include elected representatives or board members.

Also in line with the main model, there can be great value from working with others from outside, such as consultants and community representatives. They bring valuable experience, are often able to 'say the unsayable' (that which internal people dare not say) and are essential when moving to deliver on any increased capacity.

Take stock at the start

Taking the point about capacity release, and a classic diagnosis before prescription approach, a first step is to take stock of what is and what is not happening in engagement processes now, by looking at current practice in the following terms:

- Cataloguing the requirements placed on the organisation (by central government, other agencies, its own policies) to do better engagement, and any related principles and guidance.
- Cataloguing, and quickly evaluating, the various forms of consultation or engagement already under way; how much, by whom, on what, to what effect etc.
- Auditing staff (and member) skills, experience and attitudes at all levels.
- Checking, as far as one can, expenditure on engagement.
- Checking organisational barriers to and opportunities for making progress internally and externally.

The second step is to attempt to, as it were, populate all the boxes in the basic diagram (Figure 27) introduced earlier. This might involve seeking answers, from a wide range of people, to questions such as:

Culture and Direction:

- To what extent is the agenda of achieving more and better quality engagement well understood at senior levels and amongst elected members?
- If this understanding is present, how does it manifest itself?
- Is there genuine commitment to drive forward work on improving the organisation's engagement activities?
- If yes, where is this commitment set out and/or who 'owns' and leads it?
- Do the overall resources allocated to engagement (budget and people resources) match the aimed-for level of commitment?

Process Management:

- When working on engagement activities, is there regular and systematic analysis of who should be engaged, how, where and when?
- Who does the analysis and planning to ensure these aspects are integrated?
- Are the various engagement activities across the organisation planned and integrated? For example, is there:
  - A collective picture of what goes on across the organisation?
  - Joint planning of engagement activity to ensure good use of resources/ added value?
  - Sharing of the results of engagement activities?
- Are the engagement activities linked to those of other organisations such as the health service, police, voluntary and community groups?
- Is there investment in learning to enable staff to improve their work on engagement? If so, how does this manifest itself?

Delivery:

- Do all staff and members have an appropriate level of skills, knowledge and confidence to enable them to carry out, or contribute effectively to, any engagement activities relevant to their job?
- What training and experience is available to staff to enable them to continue to learn and to improve their practice?

Know the organisation

Having understood what is currently under way and the extent of genuinely corporate support for better practice, it is then important to consider the general nature of the organisation, its ethos, roles, context, size and so forth. This is because some organisations, almost regardless of their main role, can be highly technocratic while others can be administratively driven. Some may be highly professionalised, others very much determined by the political (small or large 'P') direction of the key decision-makers. Some are highly structured and rigid, others very flexible and responsive to their external context. And, of course, some are a bit of each.

This is not the place to rehearse different ways to tackle all of these options but the key point must be made: Be very clear about how change can happen in any specific type of organisation, how it might be blocked (by whom, on what basis), how in general change is best initiated and consolidated and what windows of opportunity are available (e.g. some new legislation) through which to get started.

Develop a list of 'building blocks'

Although many organisations automatically reach for classroom training as their default option, there are other ways to deliver learning. The little available evidence suggests that forms of action learning based around doing real work are most effective, and must then be consolidated through more traditional training. A range of approaches should therefore be employed to ensure that staff will actually *apply* new insights and skills. This is important for learning about engagement because this is not a purely technical or procedural topic. Other approaches can include:

- The use of 'shadowing' to pair up a less experienced staff member with an experienced staff member, providing opportunities to watch new practice and perhaps assist.
- More generally, providing experienced support for staff to work on exemplar projects, especially those that involve use of an unfamiliar, perhaps challenging approach to engagement.
- For both of the previous two points this could be arranged with an external specialist consultant, either ad hoc on a single project or through some form of placement.
- Getting close to formal training, it is possible to arrange some occasional 'hot topic' seminars. Putting on a sequence of attractive, short and accessible seminars for staff (e.g. over lunch) to come along, gain their interest and raise awareness.
- For all of these it is also important to set up an informal support network so that trainees can meet regularly to share their experiences and continue to develop their learning.

In parallel with this, it is important to put in place competency frameworks linked to personal appraisals and job descriptions because these begin to place engagement as a core skill rather than just an optional one.

In terms of more traditional training, it is essential that this is carefully planned over time; an ad hoc series of unrelated courses can be just as damaging as ad hoc engagement projects. The following are key points to consider in planning a training programme:

- The length of training courses requires careful consideration. Some staff will only be able to sign up for a half day course, yet a good full day course is probably several times more effective than a half day one. A partial solution is to split an event: Two half days or two days separated by a week, for example. There again some training, notably on facilitation skills, simply requires more time; most trainers suggest a minimum of two full days for this topic to enable people to have the opportunity to practice during the training.
- Training should also be tailored to particular issues and to the experience and skill level of the participants. Admittedly, this can make it difficult, even in larger organisations, to attract an appropriate number of trainees all at any particular level. Because engagement is a generic skill, one solution is to train different people together; a mix from various departments, perhaps even from other organisations (private, public, voluntary and community sector).
- Difficult though it is (except very occasionally) there is real value in training a whole team together.
- Training for engagement does not mean people sitting in rows being lectured. Collaborative working is a craft, so training needs to be active, interactive and practical and in itself demonstrate the best of working together.
- Where training takes place is surprisingly important. Training sessions that are held within an office setting can affect people's focus when the day job is just outside the door. Outside or semi-outside venues are good but care is needed to ensure they are appropriate for the type of training proposed. For example, a fixed-seat venue is inappropriate for active training on group work.
- Consideration should be given to who does the training. Outsiders can bring independence, high level skills, wide experience and an element of status, but may not be aware of an organisation's context or culture. Internal trainers may know the organisation really well (perhaps too well) but may not have genuinely wide experience. Internal trainers will, however, be more able to consolidate learning in the days or weeks after initial training has happened; something an external trainer would find difficult. So both may be needed.
- Finally, follow-up after training is crucial to ensure reflection with participants on how they have applied what they learnt in the training in their real work.

Tactics

Organisational capacity building for engagement is a long-term commitment. There is no simple or preferred way to go about it. A lot depends on how things work within any organisation. That means thinking tactically about how to move things forward. For example, some basic choices might be to:

- ... start slowly then move up a gear or two? It is wise to aim for modest goals in the first instance and build on success. Setting out more challenging goals will require a substantial, long-term organisational change process for which the organisation may not (yet) be ready.
- ... try to work heavily on one box from the diagram in turn, or try something for each over time? While it can be helpful to develop in-depth understanding and experience across a given cohort of staff or area of activity, this can only be taken so far. It is best to avoid doing too much in any one box because that will soon become counter-productive.
- ... do things that bring people together from different organisational levels, teams, departments, or work with each alone? Or a mix of each over time? If it is decided to work with one specific team/activity area, it will still be critical to include any other internal players who have relevant skills/roles, for example, communications officers.
- ... build from what has worked and is working, i.e. what the people at the coal-face are doing now, or start building capacity at the higher levels? Beware of going too far down any single route without securing understanding and buy-in from middle and senior management in particular or, again, there are likely to be barriers to progress.
- ... work alone or with others? This might be adjacent authorities/ departments, other local organisations, even the private sector and local communities. Many others are also being challenged to do more and better engagement and a sharing of skills and resources would save money, provide greater challenge and help to build capacity amongst those one is aiming to engage. This is very strongly recommended.

Monitor and review

There is little to say under this heading except to stress how important it is. Evaluations and reviews from specific collaborative initiatives should be standard practice. Looking at all of these to assess an overall picture is also important. Too often lessons from training are lost once people move back into the hurly-burly of their regular job, so follow-up is important. Having done an overall organisational analysis at the outset to show which of the diagram's boxes are filled, repeating this exercise quickly (e.g. after a year) can show both progress made and gaps remaining to be filled.

At this stage the Waterton example runs out of value because the council involved had no overall approach, framework or culture of engagement; the regeneration work was entirely one-off. In fact, it is difficult to find examples of large organisations attempting capacity building for engagement. Some work has been done in some central government departments and a few local authorities in the UK (see later for one example) but none of this appears to have yet made a genuine change at overall, corporate level. The only example worth describing, and still ongoing, is with the national Environment Agency (EA), although those involved (including this author) would not suggest that there is yet coherent action in all boxes as per the earlier diagram.

Environment Agency 'training'

This example is valuable because of the range of approaches that were taken. For quite a while, people at senior, middle and more junior levels within the EA had begun to consider how to move from a communications/public relations model of working with communities to a more genuinely collaborative one. To address a major national issue (coastal flooding) a pilot engagement project was run using solely outside consultants from InterAct Networks. This project was generally successful and some other successes followed, more often involving internal as well as external people. This led to a high level commitment by the EA to a programme of organisational capacity building. Next stage actions included the following (not an exhaustive list):

- An initial outcome was a policy statement/guide for EA staff (and that could also be shared with others) about the value of engagement and some of the key principles of delivering it.
- In order to ensure that staff across the EA were familiar with this guide (just printing and circulating it was not thought to be enough) a series of awareness raising events was held.
- That moved on to more formal training for often very mixed groups, for example, on process design or facilitation, and to a number of other activities that could be called mentoring.
- The aim was to establish at least one trained, knowledgeable mentor in each main area office who would support others.
- In one case, a whole team was taken through a combination of training and action learning, i.e. they learnt by dealing with specific current examples and were supported by the trainer.
- Events were held on specific issues such as recycling and cement manufacture.
- A detailed practice guide was produced (Environment Agency, undated).
- Expert advice and support was provided on specific projects.

Facilitator networks[4]

The Environment Agency example is also of interest because one significant outcome was that any member of a technical team faced with a challenging issue requiring engagement was able to call on support from one of his/her colleagues in the office. This was a version of a model developed more widely across the UK by InterAct Networks entitled 'facilitator networks'.

The aim of facilitator networks was to establish in one area, a team of people trained in all aspects of engagement processes and facilitation that could advise on, support and facilitate events. That team could meet up on occasion to share experience and further develop their skills. Those involved would ideally be from more than one organisation, certainly from more than one department or team. The idea is that by using someone from, for example, the health service, to facilitate a council recreation department event would produce a good level of independence while avoiding drawing in (and paying for) an external person. (And someone from the council might then go to facilitate a health service event.)

The idea was started in 2002 and secured good support from a number of organisations. In one case a County Council (Derbyshire) initiated the establishment of a network and subsidised the opening training. Those involved came from the County Council, District Councils, the health service, the police, government agencies, NGOs and even some community organisations. Having been fully trained, this group of around 40 people supported over 200 events in their first full year.

Though still an extremely valuable model to form part of an overall organisation (or multi-organisation) approach, the 15 or so networks that started proved difficult to sustain. People would leave their job or be promoted and in general the people involved found it difficult to find time to meet or share best practices regularly. In addition, as shown by the more successful examples, networks need at least one highly-motivated person at the heart to keep it all moving along and, as that role was never formally recognised, these key people in particular moved on (as happened in Derbyshire). Despite this, the core principles of the model, i.e. local or intra-organisational sharing of experience and mutual capacity building, still have real value.

Procedures and structures

In order to prepare and deliver a coherent approach to a medley of engagement activities, an organisation needs to have several things in place once capacity is built. The following four elements are generic; they can and should apply to all engagement or consultation initiatives over a period of time, maybe also with some form of subset for areas of work such as planning where there is a clear national legislative context.

- **Strategy**: An overall statement of policy, ambitions, budgets and so forth, and the processes to be put in place to deliver against those ambitions.
- **Guide**: An overall set of guidance principles, processes, methods, people to access etc. for any specific engagement initiative.
- **Codes of Practice**: This sets the basic standards for those within an organisation and forms the bottom line against which practice can be tested.
- **Protocols**: A set of agreed, mutual working arrangements between the core organisation (the holder of the strategy) and other key stakeholders, both generically and then for specific initiatives.

With these in place (though some can be combined) there is a need for two further elements:

- **Programme**: This probably needs to be an annual programme to cover major initiatives that can be planned that far ahead (e.g. a regular review of service delivery), providing relevant details for each initiative of purpose and scope, who will lead, timing, budget and so forth.
- **Ongoing Database**: This is a regular access point for people working inside and outside an organisation to check what initiatives have recently been completed (to see results), what are under way and perhaps what will be starting soon (to get involved).

Self-evidently, these elements will need people to set them up and manage them; namely, a **Coordinating Group**. Caution is needed here because the few examples where some or all of this is in place are larger, urban authorities. Smaller authorities or organisations would struggle to have all of what has been described in place, certainly to be able to resource its ongoing management, though this further reinforces the argument in favour of organisations working together, sharing skills and developing some form of network.

Each of these main elements is now described more fully and this section ends with an example that shows much of the following in action.

Strategies

The need for some level of coherence may be poorly covered but it is not new. It surfaced in guidance from central government in 1998 in a paper entitled 'Modernising Local Government' (Great Britain, 1998). The paper addressed a wide range of aspects, not limited to planning and environment, and within a chapter on 'Involving Local Communities', there was the following sub-section on 'Consultation Strategy':

> ... it is also important that an authority should develop a clear view on how to encourage involvement .... It should assess those matters which could benefit most from public participation, the results it hopes for, and what those factors mean for how that participation should be achieved and who should be involved.

It is worth noting that much of this picks up on aspects covered in this book in relation to specific projects. However, the paper makes clear that this also needs to be considered at an overall level, across all engagement initiatives and that this requires some form of strategy.

The paper was, however, a consultative document only. It was followed later in the same year by 'Guidance on Enhancing Public Participation in Local Government' (Great Britain, 1998/2). Chapter 3 was entitled 'Assessment: Towards a Participation Strategy' and it suggested four key components of a strategy:

- **Comprehension**: Clarity about goals, what current good practice can be built upon, how to select techniques, how to establish monitoring and review.
- **Communication**: Systems to enable people to know how to consult and what opportunities exist, using a wide range of methods.
- **Capacity**: Interestingly, given comments earlier on capacity building, this was directed partly internally but mainly externally, at citizenship education and community development.
- **Connections**: Linking people in at all levels but, in particular, ensuring that *"participation* (is not) *seen as an alternative to the representative political process"*.

Engagement (or consultation or involvement) strategies were not unknown at this time; the guidance quotes three examples. Although it never became a formal or legal requirement to have a strategy, more followed the 1998 guidance, although many were not and are not really strategies; they were simply guides on what to do, why to do it and how to do it. In that sense, Statements of Community Involvement (SCIs) – about planning in particular (see Chapter 2) – were a major step forward, if addressing planning alone. Good SCIs not only include principles, standards and practical guidance but they usually also cover programmes for involvement on plans as a whole.

There are, however, other things that ought to be in a useful strategy, and strategies are of almost no value unless followed up with an overall approach to developing and managing a programme of engagement activity. A strategy that simply outlines why, who, what and when for any project is no step forward if there are still too many and overlapping projects under way. To be effective, a strategy needs to lay down some principles and processes by which there is:

- an overview of all engagement work;
- some priority-setting between initiatives (perhaps also stopping inappropriate ones or linking some together);
- ways of coordinating to prevent overlap and repetition (of invitees, topic or date);
- consistency of standards and appropriateness to context;
- access to and the distribution of resources;
- a clear picture of roles and responsibilities;
- an audit of available skills and how to top up if necessary;
- an overall timetable with key stage points;
- regular monitoring, feedback and strategy review.

What is also clear from some of the early strategies is that they were produced by a very small group of people *for* everybody else. It ought to go without saying that, in just the same way that any specific process design should be developed collaboratively, so any strategy should be developed with very wide-ranging input.

Coordinating group

Once a strategy is prepared it does not just run itself; it needs some sort of group or team of people to manage it, monitor it, check it, make key decisions, update the strategy and so forth. Once again, collaborative principles must be quoted. Having just middle or senior officers on a group is probably not good enough; there needs to be input from the very top, and that should include at least one elected representative, and from lower ranks, the people who actually undertake the work. However, although such a group can manage strategically and might meet fairly regularly, they cannot deal with the many day-to-day issues that arise if several initiatives are under way at any one time. Some authorities employ a Consultation Coordinator (or similar title). This is a positive step forward, but such people are usually relatively junior and require good back-up from a higher level in order, for example, to stop two teams going off on their own in the same area and ensure that they bring their engagement work together to avoid waste and public confusion.

Codes of Practice

Such codes are familiar in large organisations. Though never very detailed, they set the overall baseline for good practice. In that sense they really only come into their own when they are used in tandem with strategies and guides. A code will normally cover the overall aims of a procedure or policy, its purpose, scope, principles, outcomes, means of recourse for those believing it has not been followed, monitoring and feedback. It is a statement to those inside an organisation *and* to those outside.

Guides

Strategies are and should be broad and generic. They do not explain how engagement or consultation processes should be run, that is the function of a guide and, although a guide could be incorporated into a strategy, it is probably best to keep them separate. In addition, administrations can change as can strategies, yet it is unlikely that any guide would change, except perhaps to update and refine it.

An important caution is necessary here. Most of the guides or toolkits that emerge from searching online suffer from the same weakness referred to several times already in this book. They tend to focus very heavily on methods ('ingredients') and have little or nothing to say about coherent overall processes ('recipes' or 'menus'), yet they should of course cover all of these.

Protocols

One further way to inject more coherence to overall programmes, almost essential if the push is towards collaborative working, involves the introduction of something called 'protocols'. It is all very well for an organisation to put its own house in order but a protocol can help to ensure that other key parties also bring some coherence to their own approaches. A protocol is an agreement between two or more parties about how each will behave in certain situations and especially about how each will take on some of the most important responsibilities, not leaving this all to one party alone. As such, a protocol is a complement to the Code of Practice.

For larger development projects there is already a form of nationally supported protocol in place, termed 'Planning Performance Agreements' or PPAs (Advisory Team for ATLAS, 2008). A PPA is signed by the developer and the planning authority and outlines agreements about key dates, sharing of information, mutual speed of response, requirements at application stage and so forth.

Very importantly, the missing partner in most current protocols and PPAs is the local community as a third signatory. This tripartite approach is just beginning to be used, for example, this author supported Stroud District Council, development industry representatives *and* Parish and Town Councils in producing a 'Pre-Application Community Involvement Protocol' for Stroud District (Stroud District Council, 2014). Bristol City Council already operates a protocol,[5] developed mainly between the authority and the development industry, that they try to persuade applicants to use (it is not a formal requirement). This has already been successful to the extent that even the more reluctant developers are recognising that they can get permissions quicker if they work collaboratively, while communities are also seeing better projects emerging.

A protocol in the planning world therefore lays out, as in Stroud, the agreement on what local authorities in general will do to support an engagement process, what developers/applicants will do and what the local community is expected to do (perhaps also others, such as a statutory agency). A protocol is, therefore, likely to require the following for each engagement process:

- Establishing a clear point of contact that will enable a two-way flow of information regarding the plan/project.
- Agreeing basic milestones and any specific key target dates, as well as broad agreement on turnaround times for dealing with reasonable requests.
- Supporting the identification of key stakeholders and of local communities most affected by the plan/project and helping to identify the methods appropriate for engaging them at specific stages in the agreed process.
- Providing support in communicating with key stakeholders, helping with identifying the full range of community views and then representing these to decision-makers.
- Ensuring the sourcing and use of objective information on disputed areas of debate that is reliable and independent and which conforms to appropriate national standards and/or guidelines.
- Agreement on resource provision; who pays for what and who provides what (in one case a community agreed to provide event venues).
- Making clear that anybody's engagement with these processes is in no way an indication of support for any resulting plan or application.

Programmes

With a protocol in place, the next element is a programme of engagement activities. For those authorities which produce them (still a minority), such programmes are usually produced annually then updated perhaps monthly. The best are genuinely corporate, although there is an unfortunate if almost understandable history of (land use) planning teams creating their own independent programmes. This is probably because, unlike most others, planners are legally required to do engagement so they have formally adopted principles and thus end up doing more of it.

A good programme will enable internal people to ensure that any new suggestion for a consultation process fits in with others in terms of when it happens, what it covers, who it targets and how it is implemented and for the coordinators to manage this overall (in terms of time, resources, people etc.). If prepared annually, the programme will only be able to include regular, pre-planned and probably larger consultations, so regular reviews and updates are important.

Database

All of the programme information needs to be kept on some form of internal, website-based database. For someone *inside* an organisation needing to plan a new initiative, the database can let them know whether there are other similar projects planned, under way or recently completed in their own topic or geographical area, who that project is targeting and so forth. It can also offer contact details to enable appropriate liaison and shared planning; for example, could two projects be combined or one follow on from and complement the other?

This also has real value for those *outside* the organisation. That can be an individual or small group keen to know whether there is something planned for them on their current issues of concern. A database can warn them it is coming, outline its scope, methods and timing and ensure not just that they do not miss it but also that they can contribute in the most valuable way as possible.

For each specific initiative it is valuable for a database to include the following:

- An overview of topic, geography, target groups etc.
- Why the initiative is being undertaken, its purpose and scope.
- The overall format and process design.
- Any general methods being used and any that have specific deadlines or targets (e.g. a workshop for young people).
- Start date, end date and any milestones between.
- How/where to find out more information.

Organisations and others

Thus far we have assumed that there is only one organisation in any area. This is clearly wrong; the classic example being the existence in the same area of a local authority and a health authority. One of these may have a good approach in place while the other does not. Even if both have good approaches, this can cause significant community frustration because the approaches are still unlikely to be consistent. Other examples would be a local authority and a government agency. In the Waterton example, attempts were made to bring in some engagement by a government agency, but this was not successful. It is probably far too optimistic, perhaps even impractical, to suggest that every organisation develops entirely consistent principles, approaches, codes and so forth but this book offers the way through in such circumstances and the answer is extremely simple in principle: Meet early and collaborate!

A practical example: Bristol City Council

It is probably another example of the fruitless search for absolute zero to suggest that any authority or organisation would ever put all of the aforementioned suggestions in place. To this author's knowledge, coherent, authority-wide approaches are very rare. There is, however, at least one example that comes close to what has been suggested as good practice.

The example is the work of Bristol City Council (BCC), both corporately and within the planning teams. However, as illustrated at the start of this section, even where a good corporate system is in place, there can still be some teams, or even whole departments, who go their own way: either they will fail to undertake consultation when they should or they will devise their own standards for how to do it. Despite what follows, this is occasionally still the case in Bristol. The other proviso to the Bristol work is that the overall approach delivers various forms of consultation, not the more intense, in-depth, collaborative working elaborated in this book, though that still happens on some planning projects in particular.

Bristol City Council's approach includes the following[6] (endnotes provide web links):

- **Strategy and Code of Practice**: BCC has a form of Corporate Consultation Strategy in place. They call it a 'Code of Good Practice on Consultation'[7] and it applies to the work of all departments, units and teams within the authority. It is also promoted externally to help ensure that the authority is known to be committed to thorough consultation and has consistent principles for its practice. The authority also has an extremely good Statement of Community Involvement[8] produced through a very thorough and quite challenging dialogue between council planning officers and people from local NGOs.
- **Leadership and Staffing**: BCC currently has a small team of officers to manage its approach to consultation: the Consultation Research and Intelligence Team. There is a direct link to an overall Service Director

who leads and champions consultation activity across the authority and ensures that it has a high corporate profile.
- **Guide**: The Council supports the Code with a Consultation Toolkit.[9] This covers aspects such as 'Why do we consult?', 'Who to consult?', 'What to consult on?', 'How to consult?' and it also emphasises the importance of feedback to participants.
- **Protocol**: There is, very unusually, a city-wide planning-related version of a protocol in place in Bristol (as mentioned earlier).[10] This relates very strongly to the SCI and was produced by the City Council planning team and representatives of local development interests. The protocol is for major projects and outlines several stages for pre-application community involvement, coming close to much of what is promoted in this book.
- **Programme**: The aim of the consultation team is not just to be aware of all consultations as and when they happen but also to encourage and shape them and, ideally, to avoid the sort of overlaps and duplications highlighted in the example at the start of this chapter. This generates a thorough programme of activities but there are still occasions where officers or teams do not consult when they should or consult in their own way. Most of the programme is also available through the database (see below).
- **Database**: BCC operates a website, open to all, called the 'Consultation Hub'.[11] This lists what is under way and, for each activity, something about why, what, when, who for, plus a named contact.

Picking up on the point about developing approaches with others outside the core organisation, Bristol also offers one other probably unique arrangement, one that others could well learn from.

For some years, Bristol has been developing a number of small neighbourhood-based groups whose role is to watch out for forthcoming planning and development projects, anything from the city plan to a small shop development, and then to ensure that their local community has its say at the right time and in the right way. Those groups come together in the overall Neighbourhood Planning Network.[12] This is a voluntary organisation, part-funded by the Council and it has established an extremely constructive working relationship with planning officers and councillors to the point that network members have regular opportunities to meet with the planners to get up to date on new legislation or guidance, hear about forthcoming plans and major projects ahead of time, discuss neighbourhood planning as a whole and even look at specific major scheme proposals; it is a superb example of mutual capacity building. From a developer's perspective this offers a fast-track into a community early in the process and access to people who understand much of what planning and development are about. Though some things still slip (or batter) their way through, the best of what is available in Bristol, both generally and in specific relation to planning, comes close to at least offering the potential of win/win solutions; it is well on its way to *making it mainstream*.

Key messages

- Good one-offs are good, but not good enough; there can be real losses, even problems (wasted time, wasted costs, frustration all round, always starting at the lowest point next time) if there is no coherent approach.
- In any organisation there needs to be a whole culture of engagement; it needs to be the norm, to be the default position, to be *mainstream*.
- That culture needs to be strong at all levels from top to bottom and including key decision-makers (members/board).
- That culture should ideally be shared or understood by other organisations and the wider community.
- Capacity can be built or released and some key points about achieving this are:
  - Work across the whole organisation.
  - Take stock of practice at the outset.
  - Know and work within your organisation's overall style.
  - Develop a list of building blocks for moving forward.
  - Think hard about the most appropriate tactics.
  - Be sure to monitor and review.
- A mix of on-the-job and formal training is always needed.
- A key way to start can be by setting up, formally or informally, a facilitator network.
- In order to deliver successfully, the key points are to put in place:
  - A Strategy.
  - Some form of relatively senior Steering Group.
  - Some sort of Guide.
  - Some Codes of Practice.
  - Protocols for various parties to agree to.
  - An overall Programme.
  - A regularly updated Database.
- For larger organisations a dedicated engagement team can be highly cost-effective, certainly requiring at least one person to manage on a day-to-day basis (if only part-time).

Notes

1. Some of these are not just participants; some will often run their own engagement processes.
2. Informal seminar comment by Chris Shepley, ex-Chief Planning Inspector.
3. This checklist is drawn directly from materials produced by InterAct Networks; see the section 'Facilitator Networks' later in this chapter.
4. These were developed by InterAct Networks. Go to: www.interactnetworks.co.uk
5. See endnotes 8 to 13 for all Bristol references.
6. This was the situation as of summer 2014, although BCC were at that time in the process of making some changes.
7. Go to: www.bristol.gov.uk/consultationcode
8. Go to: www.bristol.gov.uk/node/1595
9. Go to: www.bristol.gov.uk/sites/default/files/documents/council_and_democracy/consultations/Consultation%20Toolk%20v4_0.pdf
10. Go to: www.bristol.gov.uk/page/major-developments
11. Go to: https://www.citizenspace.com/bristol
12. Go to: www.bristolnpn.net

References

Advisory Team for ATLAS (2008) *Guidance Note: Implementing Planning Performance Agreements*. London: Homes and Communities Agency.

Bishop, J. (2005) *An Infrastructure for Engagement*. Unpublished. Available from the author: jeff@placestudio.com

Bull, R., Petts, J. and Evans, J. (2008) 'Social Learning from Public Engagement', *Journal of Environmental Planning and Management*, 51(5), pp. 701–716.

Colbourne, L. (2010) *Organisational Learning and Change for Public Engagement*. Unpublished, but contact this book's author at: jeff@placestudio.com

Cuthill, M. and Fien, J. (2005) 'Capacity Building: Facilitating citizen participation in local governance', *Australian Journal of Public Administration*, 64 (4), December, pp. 63–80.

Environment Agency (undated) *Building Trust with Communities*. This is mainly an internal document but a basic version can be accessed via: http://repository.tudelft.nl/assets/uuid:6c88a7b9-1478-435d-ad58-95b7b59868b2/ComCoastWP4-07.pdf

Great Britain, Department of Environment, Transport and the Regions (1998) *Modernising Local Government: Local democracy and community leadership*. London: DETR.

Great Britain, Department of Environment, Transport and the Regions (1998/2) *Guidance on Enhancing Public Participation in Local Government*. London: DETR.

Stroud District Council (2014) *Pre-Application Community Involvement Protocol*. Available at: http://www.stroud.gov.uk/info/planning/Pre_Application_Community_Involvement_Protocol.pdf

9
Conclusions ... or an Engagement Utopia?

Writing conclusions for a practical guidebook could involve nothing more than a list of platitudes and repetition of the key messages which have already been provided at the end of each chapter. Instead, a different approach is taken (based on Lynch, 1984). This chapter tells the story of the imaginary Borough of Metford, which is adjacent to Halltown District, equally imaginary. (Though imaginary, everything is in fact a combination of the best of actual proven practice and real examples.)

We will set the scene by giving some context and background, i.e. the situation a few years ago when both authorities rated poorly on anything about consultation or engagement. (*'Isn't engagement something to do with two people who are going to get married?'*, Councillor Brag.) We will then move to today when almost everything this book covers is, almost miraculously, now in place. (*'I'm pleased to tell my constituents that Metford now leads the world in engagement'*, Councillor Brag again.) The story will then illustrate how this remarkable progress has been given life with two examples of recent engagement processes, one in Metford and one in Halltown, both apparently now quite routine (*'Doesn't everybody do this?'*, Councillor Brag) in terms of current practice, if not so routine in their outcomes.

The places

Metford is a very urban area, dating mainly from the early Industrial Revolution, with almost no countryside within its borders but a lot of now derelict land. The population of around 187,000 is extremely mixed in socio-economic and cultural/ethnic terms. There are some very poor areas in the east linked to the now empty shoe-making and wire-making factories and some fairly prosperous communities to the west, many of whose residents work to the south of the authority in an area that a Metford councillor recently called *'our very own Silicone Valley'* (three metal sheds, some trees and a lake). The town (they try to call it a city) centre is struggling because of its proximity to two other larger centres and because of a sprawl of out-of-town developments to the north. Metford is a unitary authority so it controls all local authority functions.

Halltown District is adjacent to Metford to the north west and is a District within Upshire County, so each manages some but not all local government functions. Halltown includes one large town (Halltown itself) and a considerable rural hinterland with small and large villages (*'The Market Garden of the West'*: *tourist brochure*). The town is in need of serious regeneration while some of the villages are very prosperous and the locals (*or 'yokels' as those from Metford call them*) are vociferously against any development at all, anywhere near them. Most of the small number of rural residents who are not over retirement age commute to Metford (certainly not Halltown), mainly to Silicon(e) Valley.

Both Metford and Halltown are now controlled, if narrowly, by the same political party, they recently decided to try to work together more fully, share resources and skills and so forth.

Metford, Halltown and consultation five years ago

While undertaking some engagement work in Metford and Halltown five years ago, Facilitate Consultants had an opportunity to study how things operated generally on consultation and engagement issues. A leaked version of their rather damning (so also highly confidential) report to the respective Chief Executive Officers made the following comments in its Executive Summary (the main report is 175 pages long):

Executive Report from Facilitate Consultants: Consultation in the Districts of Metford and Halltown, 2005

The three quotations used below as a preface to our main conclusions are, in our view, indicative of the issues that need to be addressed:

'Now I'm elected, I don't expect to have to talk to my constituents again till the next election', Halltown councillor.

'I've done years of hard training to become a planner. What do that lot in that pensioners' village know?', Planning officer.

'We put on an exhibition just before our application – what more do they want?', Developer/applicant.

Main conclusions:

- Forms of consultation take place in both authorities on a wide variety of plans, projects and services, operating entirely independently, using varying approaches, often addressing the same people and groups and even, on occasion, taking place in the same neighbourhood at the same time.

- Neither authority was able to provide any information on how much this costs them.
- As we know from our previous work with you, neither Halltown nor Metford has any coherent structure or procedures to manage this activity.
- Metford has a Consultation Strategy in place, which is effectively just a guide to how consultation should be done, not a strategy for doing it.
- Several members and officers in both Metford and Halltown were not even aware that this was in place.
- Some of the specific consultations in the last few years have actually been good, supported by some well managed feedback from recipients.
- The Metford City Regeneration proved largely successful, as did the engagement process put in place by the Halltown planners for their new Local Plan.
- However, both authorities have received formal complaints about consultation, one of which has now reached and been upheld by the courts through Judicial Review.
- There does not appear to be any real awareness of, or commitment to, better consultation amongst councillors or chief officers in Halltown and only some in Metford (from some recently elected councillors).
- In both authorities, decisions about consultation processes are taken by middle managers but they currently feel extremely exposed with significant budget and staff cuts under way; consultation is seen by them to be a very low priority and of no value in their career development.
- The bulk of the responsibility therefore rests with a number of junior officers, many of whom are strongly committed to better consultation. They prefer the term 'engagement'. (Surprisingly, two people even used the term 'collaborative working'; have they been reading a book?)
- However, these junior officers are under stress and have no opportunity to share skills or experience and to build their own capacity and that of others.

In summary:

- From quick checking with local businesses, voluntary and community groups, the impression that both authorities give is that the few good examples are outweighed by the generally average and poor ones and neither authority is therefore regarded as good at consultation.
- Our own calculation of the total costs of consultation to both authorities is in the order of £1,400,000 per year and much of this could be saved, or at least used significantly more effectively, through proper strategies and programmes and support and training for those required to commission and deliver engagement.
- Doing this jointly between the two authorities and with business, voluntary and community sector groups would have considerable added value.

Metford, Halltown: engagement today

Since five years ago, with both authorities now under similar political control and with a whole stream of new and keen councillors elected (Councillor Brag has just retired), things have improved significantly. A joint Working Group examined the consultants' full report (after it was dragged out of the bottom of a filing cabinet and the dust was blown off) and they have now put in place a number of measures to develop consistent and strong approaches to consultation. In fact they now more usually use the term 'engagement'. Despite their original damning report, Facilitate Consultants were recently (and to their own surprise) invited back to take stock once more and report to members and officers from the two authorities.

The consultants started by observing, and talking to facilitators for, and participants of, some specific projects. They then interviewed a number of members and officers in each authority, some the same as before, some different (and quite a few new) and held four discussion/evaluation sessions with mixed groups. Once again, some key points from the Facilitate Consultants report have been published, shared openly and not leaked this time, as follows:

Executive Report from Facilitate Consultants: Consultation in the Districts of Metford and Halltown, 2010

'I was elected again because my constituents thought I'd done a really good job of talking to them and making sure their voice gets heard properly', Metford councillor.

'It's amazing how, just by engaging them properly, we reached agreement with village communities on the thorny subject of where to locate development', Planning officer.

'I'm amazed; no significant objections and our application through in lightning quick time!', Developer/applicant.

- Great strides have been made since our last appraisal, not least as a result of the close working relationship between the authorities. The way that people now act as friendly critics for each other is especially valuable, as is the fact that this clearly involves staff from all levels and even some councillors.
- We particularly liked the varied way in which capacity was built within the organisation: challenging pilot projects, some mentoring and shadowing and ensuring that competencies in engagement are now part of job descriptions and staff appraisals. *(We greatly enjoyed having two staff and one councillor joining us on one of our projects.)*

- The approach to more formal training was also innovative, running four courses (two on process design, two on facilitation) spread over three months. Most importantly, this arrangement brought together senior staff, junior staff and members as well as people from the voluntary and private sectors.
- We particularly liked the idea of working with four other Upshire Districts to deliver the training. As you were further ahead on engagement, it has helped the others to also move forward (as they are now beginning to do, if rather slowly).
- Looking ahead, we would value talking through this report to the small lunchtime 'swap-shop' groups that have now been running for some time to share experiences.
- There is now a single, common Strategy and Code of Practice and it is good to know that the Coordinating Group has already undertaken its first annual review and fine-tuned both documents.
- You are still finding it extremely difficult to work out exactly what consultation and engagement costs (not just for you but also for communities and NGOs). However, you do believe that, at the very least, you are using all resources more effectively and we believe that is probably the case.
- We see you are on your way to producing an overall Guide *(feel free to use our material)* and it is probably sensible not to rush this while you continue to build your capacity and that of others.
- It was brave of you to try to develop one common database of stakeholders, consultees etc., but also a good idea given that many have a focus on areas and communities that do not correspond with your own legal boundaries. *(A small warning – this is a never-ending task!)*
- We notice that you have chosen to operate two separate programmes. That is always a difficult decision but you have at least used the same format for both and have a mechanism in place to ensure that initiatives covering both authorities or crossing the borders get picked up on both programmes.
- We see that, for development projects, you have worked with key interests to produce a formal Protocol. Though not all applicants are yet following it, your decision to keep evaluation records will soon provide real evidence of its value. Other applicants *will* follow!
- It is really good that your training has resulted in the formation of a county-wide Facilitator Network, and really good that this also includes people from the health service, NGOs, police and even some local businesses.
- We gather that network members have already supported dozens of projects, which is excellent, as is the fact that, by working together, you felt able to fund somebody (part-time) to manage the network. *(Another small warning; good people tend to move on so we find that networks need to be refreshed every two or so years.)*

- Finally, thanks to the pilot projects, the training, the Facilitator Network and the Protocol, people in other sectors are beginning not just to wake up to the value of good engagement but also are starting to 'put their own houses in order'. *(We had a call about this just last week from the Chamber of Commerce, so that's good business for us too!)*

Giving it life

The recent changes so excited the Halltown and Metford planners that they decided to submit two of their projects for the annual national Planning Awards. What follows, minus lots of fun photos of active people, some plans and the boring background stuff, is the first draft text for their awards submission.

1. The Metford local plan

Metford's previous strategic plan – the Core Strategy – was beginning to show its age; time had elapsed, legislation had changed and so had the local social and economic circumstances (no more wire-making for one). There was a need for a new Local Plan, although this needed to build from the best of what had been done previously, not wipe the slate clean.

A Steering Group of councillors was set up and also, given the now key role of engagement, an Engagement Sub Group. Members of the area-wide Facilitator Network (in fact people from Halltown, to keep it fairly neutral) ran a workshop to develop a list of consultees and design an engagement process for the Local Plan. One outcome was the addition of five extra people, none of them councillors, to the Engagement Sub Group (for example, people from the Chamber of Commerce and the Civic Society). Though many methods were outlined in the process design, the core focus that emerged was on bringing people together. Some slack was also left in the process because everybody was sure that new issues and groups would spring up once things started (as did in fact happen).

One point from the workshop was that few people in Metford had any idea at all about the Local Plan, what it was for, its effects and the fact that they could have a say. As a result, one of the first things to do was to start some general awareness-raising (though the group knew they would have to continue to raise awareness throughout). With help from a range of local groups, the planners set up a medley of very varied activities: formal ones (mentions in community newsletters) and patently fun ones (walkabouts and quizzes).

Luckily, the editor of the local paper was keen to help, so the paper ran a regular column and a competition. The latter asked people to describe a 'Day in the Life' (a better day!) for themselves in 2025 and some wonderfully creative stories came back from all age groups. All four secondary schools also set up a joint programme to use the Local Plan as a focus for work on geography, art, mathematics, history, language and even decision-taking, and the planners, some councillors and even some local business people helped out with that.

While this was under way, officers worked hard behind the scenes to develop a very thorough, and legally accountable, consultee database. (Interestingly, three people named on old lists as group contacts had died by the time the new list was made!) They also used local contacts to develop a list of appropriate venues and newsletters etc. through which to publicise the opportunities.

The awareness-raising drifted into some early engagement through a series of very informal Neighbourhood Drop-ins. Each was promoted differently and avoided almost all mention of planning, focusing instead on what local stakeholders said were current and specifically local issues. Although this got people in the room, often up to 300 people at a drop-in over a day or so, there were still some areas where more and more carefully targeted awareness-raising work was needed. The results from all of this became a very long list of issues, aspirations, concerns and even positive suggestions. (There again, by keeping its focus open, it also attracted comments about vandalism, dog fouling and tree roots damaging houses, plus some abuse about one particular councillor.) The Local Plan website and local newsletters were used as the main ways to share this back with people, which also introduced a very interactive priority-setting or weighting opportunity to gain some idea of which issues were most important to people. As expected, different neighbourhoods had different priorities; a 'one size fits all' plan was not going to be appropriate.

The next stage was to explore all this in a little more depth, so one very large workshop was held, with just over 100 people attending for an evening, after which much smaller neighbourhood workshops were held; 25 workshops with an average of 35 people attending each. The workshops took place at times that suited each different area, some were run by planners (after basic training by local facilitators), some by Metford facilitators and the potentially more challenging ones by Halltown facilitators. The basic workshop model was generic to enable comparison of results but each event was adapted a little to its specific context. The outcomes of this stage were agreed priorities for different areas, as well as many that were common, and some ideas for what might need to be done to address the priorities, who by and where. Lots of ideas emerged that could never be incorporated into a Local Plan (one example was a visitors' guide to Metford) but these were not lost; they were passed on to the relevant authority department (and the new guide appeared just a few months later).

To prevent everything fragmenting into lots of different plans for different areas, topic workshops were held, for example, on Transport, Design and Green Infrastructure, with people from all the many neighbourhoods mixed together. Sessions were also held for specific groups, for example, local landowners and developers. This involved some of those who had been at the initial workshops but also included other people who had particular interests or expertise. Finally in this round, the planning team worked with people such as community workers to access the usual hard-to-reach groups: Young people, the elderly, disabled, students (in the evenings, when they were awake) and even those with young families (child care provided).

Things went quiet for a while for local people as the planners absorbed everything and started to pull together a first draft plan, deliberately not glossily produced and patently not complete or consistent across all issues. Before that went out to anybody the Engagement Sub Group audited it against all the very thorough drop-in and workshop reports. This produced only a few small changes because sub group members had been part of the main process and had heard everything themselves at first hand.

The challenge now was to get a rather substantial document out to, read by and responded to by the largest number and widest range of people. This is where the awareness-raising work kicked in again. There were clearly enough potentially challenging elements in the draft for every different neighbourhood, so the main tactic was again to promote to each community those issues or ideas that would most generate interest, even concern. People could then access the plan in several ways, find the bit (or rather start with the bit) most relevant to them and then comment. They could comment on their own or as a group, (there was a toolkit given to groups to help them run events to do this), they could comment via the website (or their own group's website as well), via Twitter, with a letter or even at one of the 'surgeries' (a sort of fast-track drop-in) held in each neighbourhood. Instead of planners having to defend what was in the draft at the surgeries, people came in who had contributed to it and could see, and show off to their friends about, where and how their input had been used. Local people now started to feel as though the emerging plan was at least as much theirs as the Council's!

Everything came back together through a final series of small discussion-based workshops with a slightly different mix of people to previously, partly because issues and ideas were now rather different and partly because some new groups had emerged. The very last main event brought together again the 100 or so people who had attended the opening large workshop, plus a few new people. This focused briefly on a final sign-off of the majority of the plan but real time was spent on the few remaining issues where there had not yet been full agreement. Not all of these were resolved fully, although everyone was happy for the plan to proceed. Engagement Sub Group members also came to this workshop to do a final check on the plan content and to run a quick evaluation about the process with all present.

With a few small changes the plan went out for its formal pre-submission consultation. A reasonable number of comments came back, but very dramatically fewer than for the last plan. Almost all of the comments came from people who had been invited to join the collaborative process but had chosen not to do so. This group was rather isolated though clearly, if they made significant points, those were addressed properly.

The Engagement Sub Group then received from the planners, discussed, amended and then formally endorsed the overall Statement of Community Involvement to accompany the final plan (amended very minimally following its pre-submission stage) when submitted for examination. The previous plan had received 96 strategic objections but this new plan received only one (from a national developer who had not joined the process, despite an invitation). There were also around 80 specific objections from those living by specific sites, all easily dealt with by letter. On balance, a very good price to pay for all the awareness-raising work. As a result, the examination hearing was far shorter than anything previously and the plan was approved and adopted in double-quick time.

2. Halltown Wharf

A few years ago much of Halltown Wharf was seriously run down but it was in a very high profile riverside location. There were still several river-related businesses operating from the site and wishing to continue. The site owners wanted to do some significant regeneration work. Their initial ideas included high quality premises for the boatyard businesses, new housing (mainly to generate funding for other uses) and maybe some shops, a café, restaurant etc.; all to bring some real life back to an important waterfront area.

The owners contacted the planners who highlighted that the site was really just a long cul-de-sac and it would be better to have uses requiring limited car access; maybe a residential care home, probably not shops. They also alerted the owners to Halltown District Council's 'Pre-application Community Involvement Protocol', though the owners (who lived locally) had a vague idea about this already.

After some local networking, especially with Halltown councillors, the owners were directed to a small consultancy called Collaboration Works, who design and lead engagement processes. Two of the partners in the company already knew of the good practice in Metford and that this applied also in Halltown, so they were familiar with, and fully supported, the overall Strategy and Code of Practice. (One of the partners was a sailor so she had already used the Wharf area for boat storage and repair.) Collaboration Works were appointed to run the engagement, although they said that they would wish to use some of the skilled local people to help them, especially from the area's Facilitator Network, thus helping to further build local capacity.

In order to generate funds to pay for the boatyard regeneration, it seemed best to develop some market housing as Phase 1, but the planners insisted on first seeing an overall masterplan for the whole site. The owners brought in a private housing developer (Brindle Homes) to work with them on a Joint Venture for Phase 1; no problem because Brindle had already worked with Collaboration Works on other projects in the area. Architects, landscape designers and others were also appointed at this time, all on the basis that they would also join in fully with the engagement process.

Collaboration Works (CW) started by contacting various local people and groups listed on a (rather old) database kept by the Town Council, building up an initial list of key stakeholders and known and likely issues, checking this back with everyone as they proceeded. In line with the Protocol they then ran a process design workshop with those key stakeholders and the planners. That generated an agreed basic engagement process. It was only described as 'basic' because it was anticipated that it would, and should, change as work proceeded; familiar territory in Halltown where new protest groups pop up as soon as one starts anything! The workshop also added to, in fact, doubled, the initial list of stakeholders and consultees but, to balance this, it also highlighted all sorts of people and groups who could help to reach out to the wider community. The final outcome of this first workshop was the formation of a small Engagement Group of local people and a planning officer. That group met at key stages during the process.

A key opportunity arose with the local secondary schools (a Deputy Head was at the process workshop). The schools were desperate to find real issues to tackle within their citizenship work and they felt that Halltown Wharf's development was the perfect opportunity for them. A small group of students, varying in ages, from the two schools were introduced to the project and given some very basic facilitation skills training by people from the Facilitator Network. They then went out to all their friends (physically and, inevitably, via social media channels etc.), to parents and, without even being asked, to the five primary schools to raise awareness of the project and seek issues and ideas. Add in mentions in newsletters, local papers, local radio etc. and it would have been hard to find many in the town who were not at least basically aware of the project.

The first main engagement event (CW called it an 'in-depth' event) was a stakeholder workshop. Around 30 people came along to spend an evening working with each other and alongside the owners, developers and architects. They were not there to respond to anything, except some initial site analysis work, but to raise issues and ideas for the masterplan and to agree some of the top ones; in effect, helping to set a 'community brief'. Though run by invitation only, the whole event was streamed live via the project website so that many others could see and then comment, which some did (some constructively, others definitely not).

The Engagement Group then discussed how best to plan and promote the next event, what CW called a drop-in (an 'in-breadth' event). That took place a few weeks later in a town centre church hall over a Friday (evening included) and Saturday. All sorts of invitations went out, formally (to all households) and informally (via stakeholders and their groups). Over 300 people called in to the event, many saying they could only spare a few minutes but the average turned out to be almost an hour. The range of people was good (if a little short on young families) and the geographic coverage (checked via a map and sticky dots) was also good.

At the drop-in, people could expand the list of issues and ideas, add moveable notes, place a score on some scales, choose their preferred precedent images or, if they wished, join a small workshop with the architects, managed by someone from the Facilitator Network. Three 90-minute workshops were held, with over 50 people in total attending them. Everybody who signed in (legibly) received a short summary afterwards and could access the full report, i.e. every word they wrote (which surprised some), via the website.

After this first round, the Engagement Group met to audit the outcomes. They felt that good progress had been made, that some further work was needed, specifically with site neighbours and young families, and that some of the emerging issues warranted some awareness raising further afield than just Halltown, i.e. to Metford and even beyond.

A Neighbours Group was promptly set up by CW and their specific concerns raised and discussed; much of that related to construction period impacts. Accessing the young families proved more difficult. One partial solution to this was to set up a stall at a Family Fair being held at one of the primary schools; taking the engagement to them, as it were. Given that childcare was readily available at the fair a lot of input was gained from this key group (if not quite enough, as it was at only one of the five primary schools).

By this stage the designers were producing some initial sketches. Collaboration Works came to all the team meetings and kept a running audit to show where those designs were, and were not, meeting the community brief. When it seemed that some key points had fallen into place it was time for the second round of key engagement, effectively a repeat of the first round with a workshop and a drop-in.

This time the stakeholder group had grown to nearly 45 people; some who had not seemed interested first time round now were, new ones had been added to the list and three ceased to be involved. The workshop, again streamed live, was very simple; working directly with the architects and others, small groups did their own full audit of the emerging plans against the community brief and then suggested, by drawing on plans on some occasions, further ideas and ways to tackle outstanding issues or challenges. A final session drew together group work results into some generally agreed conclusions.

The second drop-in was similar to the first, with a few exceptions. It now started on a Thursday evening (Friday had not been popular) and a larger space was used. A live Twitter feed was maintained throughout because all the drawings had just been put on the project website. That attracted lots of comments (some useful, some not, just as before). Finally, rather than small workshops, design team members, developers and owners ran short 'surgery' sessions (again supported by local facilitators) for small groups at regular intervals to give more people the chance to talk directly. Around 80 people joined some form of surgery and 420 in total passed through (lots more than the first time), again spending almost an hour on average.

The material presented at the second drop-in was also different. This time, there were definitely proposals to respond to that were getting near to being final and which had been amended following the stakeholder workshop. These were carefully drawn not to seem too 'final' and always accompanied by some 'you said, we did' notes to show how the scheme had responded to everybody's previous local input. People were asked to fill in a very simple questionnaire on a series of different aspects: movement, materials, colours, landscaping etc. The questionnaire also asked for their views on how well they felt they had or had not been listened to and whether the designs reflected that.

After all this, CW went away to produce a very thorough Report of Community Involvement and shared this with the Engagement Group, who formally endorsed it. An Outline Planning Application was submitted for the overall masterplan and a Full application for Phase 1, accompanied by the report and all the many other technical documents. Because of the extensive engagement work undertaken, there was not a single objection that could be judged to be significant (a couple of niggly ones about minor footpath rerouting, bollards and seagull damage). As a result, the application could be determined by the case officer, who had been fully involved throughout, on delegated powers and it was approved in record time. It was the largest project in Halltown ever to be approved on delegated powers.

And now ... action!

Although all the individual components of the Metford and Halltown story are drawn from real life and proven examples, the story as a whole had to be imaginary because there is almost certainly nowhere where all of these have been brought together in this holistic, coherent and cumulative way, i.e. mainstream.

This book has focused heavily on how to put in place high quality engagement or collaborative practice on specific plans and projects. Making that happen more often, even more often than not, would be a significant step forward. The hope is that this book will help with that.

The real aim, however, must be to reach the situation, as imagined for Metford and Halltown, in which working together, working collaboratively and engaging widely become the standard, default mode for everybody in every situation where that is clearly required and to a degree and in a manner which is appropriate. That would be more than a step forward; that would be a giant leap. But it is still worth hoping for.

Reference

Lynch, K. (1984) *Good City Form*. Cambridge, Massachusetts: MIT Press.

Appendix 1
Collaborative Working

It was a bold, some might say reckless, decision to include the term 'collaborative planning' in the title of this book. In everyday conversations with local people, professionals and politicians, the methods and benefits of working collaboratively (or just working together) can be discussed quite simply and usually receive a positive response. However, within academic circles in particular, every aspect of collaborative working, deliberative practice, dialogue-based processes, consensus building, facilitator independence and so forth is argued over ferociously and protected vigorously, often with as much territorial defensiveness as sometimes displayed by traditional professions. Although the defensiveness may sometimes feel a little over-heavy, there is certainly some legitimacy in the concerns about how collaborative planning gets taken up and used beyond the charmed circle of academia.

As has become clear (hopefully), even the more demanding processes and terms such as 'engagement' that are clearly a major step up from 'consultation' do not do justice to what is at the core of this book. Even 'engagement' implies a process still largely set up by one organisation and that organisation then has significant control over who is involved, the scope of discussion and the extent to which any input by others is valued and used. Though collaborative working in real, practical situations rarely achieves the ideal – complete equity between all involved – developing processes that at least aim towards that ideal is what this book has been about. So collaborative planning it is and, despite pressures to the contrary, that term has remained in this book's title.

For those interested in digging deeper, four books in particular are now well-thumbed, regular sources of ideas and challenges for this author (and between them reference dozens of others). They are:

- Acland, A. (1990) *A Sudden Outbreak of Commonsense*. London: Hutchinson.
- Forester, J. (1999) *The Deliberative Practitioner*. Cambridge, Massachusetts: MIT Press.
- Healey, P. (1997) *Collaborative Planning*. London: Macmillan Press.
- Innes, J. and Booher, D. (2010) *Planning with Complexity*. Abingdon, Oxon.: Routledge.

The two books by Healey and Forester are very much about planning in the land use related sense covered by this book, although Forester's prime context is the USA. The Innes and Booher book is about much broader issues of public policy-making and is also primarily rooted in a USA context. Each of these books can be described as theoretical although they also, especially Forester's, draw heavily on practical examples. Acland's book is about the background to, challenges to and solutions for building consensus generally. What follows draws heavily from these books and is a summary only of key points; the best way to dig deeper is to read them.

Why might another approach be needed?

All four authors start by addressing what they believe to be inadequate or inappropriate about the way in which most public policy decisions are currently made. Perhaps the most commonly used criticism of the way things work now is that they are adversarial, the very opposite of collaborative; everything is win/lose rather than win/win. Innes and Booher elaborate this, and the reasons behind it, with the following long list of key factors that get in the way of better working (abstracted for use here):

- Unfamiliarity (for all) with different ways of working.
- Lack of shared knowledge and understanding between parties.
- 'Tribalism' (another way of expressing professional territorialism).
- Increased levels of community organising.
- Government structures, bureaucracies and hierarchies.
- Elected representatives.
- Declining confidence and trust in government.
- Total reliance by some on scientific knowledge and methods and their related professionals.
- Single interest lobbying.
- A focus on parts of problems rather than on wholes.
- Linear approaches to reaching decisions.
- Legalism and over-reliance on basic law.

The authors also add that many people make the assumption that consultation means public meetings and, of course, nobody wants those!

Much of this has already been picked up in the early chapters of this book but it is chilling to see it all listed so bluntly. And the fact that, as the authors say, there is no simple formula for collaborative processes, only serves to help explain why people still default to apparently simple answers and methods, even though they often know them to be inadequate.

Responses to poor current practice

Having read the above, what can counter such a long list of criticisms, and hence challenges? What is it that collaborative approaches can bring by way of an alternative? Healey captures it well in what might be considered a definition of collaborative working:

> ... planning processes need to work in a way which interrelate technical and experiential knowledge and reasoning, which can cope with a rich array of values, penetrating all aspects of the activity, and which involve active collaboration between experts and officials in governance agencies and all those with a claim for attention arising from the experience of co-existence in shared places.

However, there is no point in pretending that collaborative working alone, as defined here, can tackle all the issues raised. For example, it can at best only contribute minimally to re-establishing trust in government, but that is at least a positive contribution. Healey's definition says nothing about unfamiliarity with different ways of working but that can be addressed, as this book has highlighted, in a number of ways. To counter this, a key point about collaborative working is its ability to address many challenges in one go, to not be bounded by one person's or one profession's terms and definitions and hence be able to deal with almost any issue that is raised.

Healey's core definition can be elaborated by a series of quotations from her and the three other chosen key authors as follows (with a rather basic one to begin with):

- "When it comes to the crunch ... it [standard practice] is really only one remove from the law of the jungle – the winner takes all and the loser gets eaten." (Acland)
- "Planners need to be aware of a double client, an employer or 'customer' for the planner's services and, more broadly, the citizens affected by the 'direct client's' proposals." (Healey)
- "A process is collaboratively rational to the extent that all the affected interests jointly engage in face-to-face dialogue, bringing their various perspectives to the table to deliberate on the problems they face together." (Innes and Booher)
- "A progressive planner has the responsibility for addressing ... distorted communications to help equalize power among the people." (Forester)
- "Public officials .... harbor doubts about the legitimacy of any sort of public decision making other than representative government." (Innes and Booher)

- "We cannot resolve questions about collective action by using the language of any one form of reasoning." (Healey)
- "Because local politicians may interest themselves in many of these demands (affordable housing etc.) planners have to work with many of these demanding parties at the same time – parties whose mutual distrust and strategic posturing regularly undermine their collaborative problem solving." (Forester)
- "Unless participants learn how to build consensus across their differences, agreements about policy directions will not endure, disintegrating at every challenge." (Healey)
- "Mediation is an art, a craft and a science." (Acland)
- "As they try to see past problems and future opportunities through the eyes of many different actors, planning analysts try to build critically informed but pragmatically viable agreements. Working in-between many affected parties or stakeholders … planners and policy analysts face a pressing and central challenge of democratic politics: the challenge of 'making public deliberation work'." (Forester)
- Spatial processes which seek consultation and collaboration are, in present conditions, unavoidably multicultural." (Healey)

Ways forward

Though mainly arguing theory, all four authors offer some more practical approaches that can be put in place to address all of the above. Once again, via quotes from all four:

- "Planning work is not just about the substance or specific contents of issues … It is also about how issues are discussed and how problems are defined and strategies to address them are articulated." (Healey)
- "Every participant must both know his or her interests and explain and stand up for them." (Innes and Booher)
- "Facilitation enables people to learn more about their situation and that of their opponents and, as a result, reach agreements which are positive and realistic for all concerned." (Acland)
- "All must have equal access to the relevant information and an equal ability to speak and be listened to. Nothing is taken for granted and nothing is off the table." (Innes and Booher)
- "If mediators are to foster plurality, foster the virtues of a rich … politics, they must be able to anticipate and somehow respond to the play of power." (Forester)

- "The giving of rights to be heard goes with the responsibility to listen, to give respect, and to learn, through procedures which foster respectful mutual learning about the concerns of others, and which draw on the knowledgeability of all members of a political community." (Healey)
- "A collaborative process needs leadership to get started. Someone has to have the idea and the ability to engage other leaders in designing and setting up the process." (Innes and Booher)
- "In practice settings, listening carefully enables not just some vague 'better communication' but a far more important recognition of what others really value, beyond their wants of the moment." (Forester)

To once again stress the ambition of collaborative working, and the strength of societal resistance to it, Innes and Booher add the cautionary comment – argued throughout this book – that all of this outlines *"conditions to be aimed at, though they can never be completely achieved"*!

Appendix 2
The Legal and Quasi-Legal Context

Please note: The material in this section is not authoritative. It is the outcome of this author's experience with legal processes, although some important and more thorough references are provided. Website references are provided within the text. A number of web links are provided as endnotes.

There are two main dimensions to legal requirements and standards set for consultation (a deliberately chosen word) in and around planning issues in their broadest sense:

- Standards set nationally and on some occasions locally (e.g. by a local authority) that are *generic*, applying to any and all situations, including planning.
- Standards set internationally, nationally and on some occasions locally (e.g. by a local authority) that are *specific to planning* and also to related areas of work.

It could be argued that there is a level above this of what is sometimes termed 'natural justice' which is therefore international. However, even allowing for international conventions on human rights, the variations between countries in terms of culture, political structures, precedent, legal systems and so forth are so considerable that this probably has little meaning unless translated, through national policy and precedent, into approaches and standards on natural justice for a specific country (as some do but others do not).

The key point to make here is that, at present, all of these are effectively standards about relatively basic *consultation*, mostly what is termed *formal consultation*, and they have not yet come fully to terms with the greater complexity inherent in more *collaborative or engagement* approaches or what some term *informal* consultation.

The other inevitable caution at this point is that much of what follows, notably in relation to planning, is specific just to England, no longer to all of the UK.

Generic requirements and standards

In 2012 the Cabinet Office updated its 'Consultation Principles':[1]

This paper *"sets out the principles that Government departments and other public bodies should adopt for engaging stakeholders when developing policy and legislation"*. It states that:

> The governing principle is proportionality of the type and scale of consultation to the potential impacts of the proposal or decision being taken, and thought should be given to achieving real engagement rather than following bureaucratic process. Consultation is part of wider engagement and whether and how to consult will in part depend on the wider scheme of engagement.

Although there is no apparent source of clarification or guidance on what is meant by *"real engagement"* or how consultation relates to *"wider engagement"*, it appears that there are three key principles, all encouraging for this book:

- **Timing of consultation:** "Engagement should begin early in policy development when the policy is still under consideration and views can genuinely be taken into account. There are several stages of policy development, and it may be appropriate to engage in different ways at different stages". It is suggested that the time allowed for responses "might typically vary between two and 12 weeks".
- **Making information useful and accessible:** "Policy makers should think carefully about who needs to be consulted and ensure the consultation captures the full range of stakeholders affected. Information should be disseminated and presented in a way likely to be accessible and useful to the stakeholders with a substantial interest in the subject matter. The choice of the form of consultation will largely depend on the issues under consideration, who needs to be consulted, and the available time and resources".
- **Transparency and feedback:** "The objectives of the consultation process should be clear. To avoid creating unrealistic expectations, any aspects of the proposal that have already been finalised and will not be subject to change should be clearly stated. Sufficient information should be made available to stakeholders to enable them to make informed comments". (There is, however, no clear mention of feedback to consultees.)

At national level there is also 'The Compact',[2] almost a form of protocol between the government and 'civil society organisations' (or CSOs, i.e. NGOs).

Appendix 2: The Legal and Quasi-Legal Context 219

This includes a number of standards and mutual responsibilities related to consultation, mainly between the government and the organisations but also treated as applying to consultation with all of society. Some key extracts include:

2.3 Work with CSOs from the earliest possible stage to design policies, programmes and services. Ensure those likely to have a view are involved from the start and remove barriers that may prevent organisations contributing.

2.4 Give early notice of forthcoming consultations, where possible, allowing enough time for CSOs to involve their service users, beneficiaries, members, volunteers and trustees in preparing responses. Where it is appropriate, and enables meaningful engagement, conduct 12-week formal written consultations, with clear explanations and rationale for shorter time-frames or a more informal approach.

2.5 Consider providing feedback (for example, through an overall government response) to explain how respondents have influenced the design and development of policies, programmes and public services, including where respondents' views have not been acted upon.

Perhaps the most commonly quoted source of clarification on standards and requirements, including for within planning, is informally termed the 'Sedley criteria',[3] named after Stephen Sedley QC. The requirements he has suggested for acceptable consultation are as follows:

- that consultation is undertaken when the proposals are still in a formative stage;
- that adequate information is given to enable consultees properly to respond (this in turn may require that there is an actual proposal in existence upon which consultation takes place);
- that adequate time is provided in which to respond; and
- that the decision-maker gives conscientious consideration to the response to the consultation.

In a paper by Jonathan Auburn QC (Auburn, J., undated, see also Fordham, 2007) drawing from case law in areas that include Judicial Review, the author expands on the Sedley criteria, queries some and adds others. In a brief summary he highlights the following general points:

- Where the duty to consult is enshrined in legislation, the legislation most commonly lists those bodies and organisations that must be consulted.

- Auburn adds that *"those not listed are unlikely to have a right to be consulted"*, however, some legislation enables those managing a consultation process to also add any others it considers appropriate.
- When statements mention 'the public' this does not require that every person is consulted.
- In the absence of a specific legislative duty, Auburn believes that it is unclear as to whether any general duty to consult can be argued, for example, on grounds of natural justice or procedural fairness.

Auburn then offers his summary views on the content of any duty as follows (shortened here):

- **Timing:** "Consultation must be at a time when proposals are still only at a formative stage, when the views of the decision-maker are only tentative or provisional upon the outcome of the consultation process and the decision-maker has not yet fixed upon a definite solution but is prepared to change course if persuaded to do so." A decision can be challenged if consultation starts after "a formative stage".
- **Contacting consultees, notification and publicity:** This "requires the taking of positive steps to make the opportunity to make representations known to those who have a right to be consulted". Auburn then refers back to the Cabinet Office standards (above) in terms of notification needing to be both thorough and proactive.
- **Manner of consultation:** This does not have to be face-to-face with every affected person, nor does it always have to include (as some have suggested) a basic opportunity for written responses. A range of methods is encouraged and, of relevance to the arguments in this book, "the court will look to whether there has been adequate consultation as a whole, not whether each form of consultation was adequate of itself"; i.e. at overall process.
- **Consultees who do not speak English:** It has to be shown that appropriate methods were put in place, via written translations or translators available at events for those unable to speak, read or write in English.
- **Consultation on options:** Although, in principle, there is no need to consult on anything other than a preferred option, this can be challenged on the grounds of unfairness if alternatives were considered internally and ruled out summarily, or if the exclusion of some options leads to the exclusion of some people who are affected by that exclusion.
- **Sufficient information:** "Consultees must be given sufficient information to enable them properly to understand the proposal and respond to it." This may need to include any or all of the following: Description of the proposed action and the information on which it is based, key reasons and

assumptions behind the proposal, the criteria used to develop and assess the proposal and (as above) information on possible alternatives.

- **Accuracy of information:** Although there is a responsibility on consultees to check the correctness of information, if incorrect information can be shown to have been used, a final decision based on that information would be unsound.
- **Existence of a proposal:** At some stage in any consultation process a clear and explicit proposal must be presented, to which consultees can make focused responses.
- **Adequate time:** This is highly sensitive to varying situations. However, even on very minor matters, a period of one month appears to be a minimum. Times can be more minimal if an issue is already well known to all or is almost repeating what has happened before. It can also be short if there is a defensible reason for urgency. In general, a period of 12 weeks is required for a conventional approach (i.e. written responses from all consultees).
- **Genuine and conscientious consideration:** "The decision-maker must give genuine and conscientious consideration to the representations received. The product of the consultation exercise must be taken into account in finalising any proposals. The decision-maker must embark on the consultation process prepared to change course if persuaded by that consultation process to do so." Judgements on this issue are often made on the basis of a report by an officer of the authority commissioning the consultation, in which case "the material provided to the decision-maker must be a fair summary which grapples with the points of substance in the objections made".[4]
- **Beyond consideration:** Very importantly for this book, Auburn states that "there is no requirement to achieve a consensus or work towards one". Achieving consensus is therefore a best practice approach, not one that simply clears any basic legal requirements.
- **Consultation on amended proposals:** Case law here is unclear, mainly because of debates about what constitutes a minor amendment, a major one or a fundamentally new proposal. If the latter, further consultation is necessary. If one of the former, and the other versions were considered properly at options stage, no further consultation may be necessary.

As outlined in Chapter 8, some local authorities and agencies have put in place (whether with formal endorsement or not is often unclear) their own **Codes** or **Protocols** that set standards or requirements for all consultation within their area of influence. Few include standards entirely consistent with those described above and some add further ones. In the apparent absence of examples of challenge, it is unclear what legal status can be given to any additional or higher standards or what rights unhappy consultees might have if they judge a process to fall by national standards not included in local codes or protocols.

Planning requirements and standards

Almost from its outset, in the years immediately following the Second World War, the UK planning system has included some rights for people to be consulted. This was, however, minimal and focused, relating to neighbour notification of planning applications. This gave a right to support or object to an application and to have any comment dealt with openly in an officer's report. This right still exists although the general view is that different authorities interpret the term 'neighbour' very differently (some very narrowly, some very widely) and consumer/community groups often express concern that, in their view, objections rarely have any effect at all on final decisions.

International law has a bearing on all planning processes, from strategic planning through to neighbourhood planning and development management. This covers the application of UN Conventions (e.g. on human rights or equalities) and the application of European law, notably the Aarhus Convention[5] and Strategic Environmental Assessment.[6] Though mainly technical, all of these have implications for community engagement. Environmental Impact Assessment[7] is an interesting example because, although the law requires consultation, it is so rare that a student wishing to do a higher degree on the topic was forced to stop because she could not find enough examples to enable a proper comparison.

Although there have been many minor and major planning Acts over the years, much legislation is simply moved forward from a previous regime, some is amended only minimally and new material is added. On that basis the most relevant Acts (as of autumn 2014) are:

- Planning and Compulsory Purchase Act 2004[8]
- Planning and Compulsory Purchase Act 2008[9]
- The Localism Act 2011[10]
- Amended Regulations produced in 2012[11]
- National Planning Policy Framework[12]

As these were all in place at the time of writing this book, some information has been introduced earlier, so all will be covered briefly and in relation solely to consultation, with fuller references given where appropriate.

Appendix 2: The Legal and Quasi-Legal Context 223

The Planning Acts of 2004 and 2008, and subsequent enabling legislation, regulations and guidance, introduced or consolidated the following of relevance here:

- A requirement on planning authorities to produce Statements of Community Involvement.
- A requirement for community involvement in strategic plan-making.
- Encouragement (not requirement) for applicants to undertake pre-application community involvement.
- A list of statutory consultees (to which authorities could add others).
- The establishment of the Infrastructure Planning Commission (IPC) with a requirement to consult on major projects.

For **plan-making** activity, the statutory consultees are now called 'general consultation bodies' and are defined in the Regulations[13] as:

- voluntary bodies some or all of whose activities benefit any part of the local planning authority's area;
- bodies which represent the interests of different racial, ethnic or national groups in the local planning authority's area;
- bodies which represent the interests of different religious groups in the local planning authority's area;
- bodies which represent the interests of disabled persons in the local planning authority's area;
- bodies which represent the interests of persons carrying on business in the local planning authority's area.

For **development management**, the term 'statutory consultees' is still used.[14] The list includes the following:

- The Canals and Rivers Trust
- Coal Authority
- Crown Estates Commissioners
- Department for Culture, Media and Sport
- Department of Energy and Climate Change
- Department for Environment, Food and Rural Affairs
- Department for Transport (administered in practice by the Highways Agency)
- Environment Agency
- English Heritage
- Forestry Commission
- Garden History Society

- Health and Safety Executive
- Highways Agency
- Local Planning Authorities
- Local Highway Authority
- County Planning Authorities
- The Greater London Authority
- Natural England
- National Parks Authorities
- Parish Councils
- Rail Network Operators
- Sport England
- Theatres Trust
- Toll Road Concessionaries

The Localism Act 2011 updated, sharpened and clarified some previous legislation but also introduced some significant new measures. The ideology of localism, and what was termed the 'Big Society', was underpinned by a commitment to giving local people a greater say over the shaping of their own local environments, so the Act was implicitly about forms of engagement or consultation. The main elements related to land use planning were:

- Updating the guidance on probity for elected representatives, especially on the issue of predetermination.[15]
- Introducing a new Duty to Cooperate between adjacent local authorities in plan-making; though it remains unclear whether or how this affects consultation.
- Introducing Neighbourhood Development Plans to be done by communities themselves (if with authority and perhaps consultant support). By definition, this ought to require high levels of engagement but that is not currently a basic condition so has, on occasion, been overlooked by Examiners (and see below for referendums).
- Introducing Neighbourhood Development Orders that enable communities to take control of specific planning applications or groups of them. This too would seem to require high levels of engagement (again see below for referendums).
- Introducing Community Right to Build schemes, enabling communities to undertake and give permission for their own developments. As above in terms of engagement.
- Introducing referendums on a range of issues (not just planning) but of particular relevance as the final means of judging community support for Neighbourhood Development Plans, Orders and Right to Build projects.

- Introducing a right for communities to limit the sale of, and bid for ownership of, Assets of Community Value (e.g. a local shop or open space).
- Mention is made of a forthcoming legal requirement to put in place pre-application community engagement. Other material suggests that this would only be for larger projects but, as of autumn 2014, the necessary secondary legislation is not in place.
- Changing the management of major infrastructure projects, removing the IPC and passing responsibility to the Planning Inspectorate. This does not, however, appear to have altered the requirements for consultation.

Serious questions remain about the use of referendums rather than relying on very thorough engagement work and results (referendums can inhibit early engagement) and case law is as yet unresolved on several issues around Neighbourhood Development Plans in particular (e.g. to what extent a local authority can influence the boundaries of area designation). It is also interesting to note that, despite the government rhetoric about greater community involvement, and some serious criticism of involvement practice following the 2004 Act (Great Britain, 2007), there was nothing at all in the Localism Act about improving practice in strategic plan-making.

The Neighbourhood Planning (General) Regulations 2012[16] introduced yet another definition of consultees. In addition to those listed above as 'general consultation bodies' for statutory plan-making, these Regulations, now using the slightly reduced term 'consultation bodies', added the following, rather similar to the list for development management:

- where the local planning authority is a London borough council, the Mayor of London;
- a local planning authority, county council or parish council any part of whose area is in or adjoins the area of the local planning authority;
- the Coal Authority;
- the Homes and Communities Agency;
- Natural England;
- the Environment Agency;
- the Historic Buildings and Monuments Commission for England (known as English Heritage);
- Network Rail Infrastructure Limited;
- the Highways Agency;
- the Marine Management Organisation;
- any person –
    - to whom the electronic communications code applies by virtue of a direction given under section 106(3)(a) of the Communications Act 2003; and

- who owns or controls electronic communications apparatus situated in any part of the area of the local planning authority;
* where it exercises functions in any part of the neighbourhood are –
  - a Primary Care Trust established under section 18 of the National Health Service Act 2006 or continued in existence by virtue of that section;
  - a person to whom a licence has been granted under section 6(1)(b) and (c) of the Electricity Act 1989(8);
  - a person to whom a licence has been granted under section 7(2) of the Gas Act 1986(9);
  - a sewerage undertaker; and
  - a water undertaker.

At the time of writing, the National Planning Policy Framework has uncertain legal status. Government ministers and civil servants have been known to state that it is law but others disagree. This is important because it includes strong statements about the importance of good community involvement. Paragraph 66 is especially significant in stating, with reference to planning applications, that *"proposals that can demonstrate this* (good pre-application consultation) *in developing the design of the new development should be looked on more favourably"*. The precise, and hence legally defensible, meaning of *"looked on more favourably"* (or its opposite of unfavourably) is particularly unclear.

Notes

1. Go to: https://www.gov.uk/government/publications/consultation-principles-guidance
2. Go to: https://www.gov.uk/government/uploads/system/uploads/attachment_data/file/61169/The_20Compact.pdf
3. The legal reference is: *"The submissions by Stephen Sedley QC were made in the case of R v Brent London Borough Council, ex parte Gunning (1986) 84 LGR 168 and set out by Hodgson at 189"*.
4. The approach to final reports taken in Chapter 7 is an advance on this.
5. Go to: http://ec.europa.eu/environment/aarhus
6. Go to: http://ec.europa.eu/environment/eia/sea-legalcontext.htm
7. Go to: http://www.legislation.gov.uk/uksi/1999/293/contents/made is
8. Go to: http://www.legislation.gov.uk/ukpga/2004/5/introduction
9. Go to: https://www.gov.uk/government/publications/planning-act-2008-guidance-related-to-procedures-for-compulsory-acquisition
10. Go to: http://www.legislation.gov.uk/ukpga/2011/20/contents/enacted. The Localism Act includes much, not covered here, that is unrelated to planning.
11. Go to: http://www.legislation.gov.uk/uksi/2012/767/made
12. Go to: https://www.gov.uk/government/publications/national-planning-policy-framework--2
13. Go to: http://www.legislation.gov.uk/uksi/2012/767/part/1/made

14 Go to: http://planningguidance.planningportal.gov.uk/blog/guidance/consultation-and-pre-decision-matters/table-2-statutory-consultees-on-applications-for-planning-permission-and-heritage-applications/
15 For guidance on this go to: http://www.pas.gov.uk/documents/332612/1099271/Probity+in+planning+guide/c2463914-db11-4321-8d38-be54c188abbe
16 Go to: http://www.legislation.gov.uk/uksi/2012/637/schedule/1/made

References

Auburn, J., undated. *Consultation*, Paper to a Judicial Review Conference, London: Thirty Nine Essex Street Chambers.

Fordham, M., 2007. *Advising in Consultation*, Paper to the Alba Conference, London: Blackstone Chambers.

Great Britain, Department of Communities and Local Government, 2007. *Lessons Report 3: Participation and Policy Integration in Planning*, London: Department of Communities and Local Government.

Appendix 3
Resourcing Engagement

What is the cost of good engagement? There is no simple answer to this question because, as should be obvious from all that precedes this Appendix, every situation is different. Is it worth the money? Once again there is no simple answer about the cost benefit of engagement, although a colleague of this author[1] has often been heard to say that *"if you think consensus is expensive, try conflict!"*

This is important because engagement advocates are often asked why there is any need to do engagement at all, as if conventional procedures have no negative cost implications. They do, because of all the minor or major conflicts that occur and have to be addressed by a local authority when trying to complete a plan or a developer when trying to secure a planning permission. A local authority failing to do any proper engagement may well have to go through an immensely expensive examination in public, yet there are examples (some outlined in this book) of proper engagement enabling examinations to be quicker and cheaper if objections are resolved earlier. A developer failing to show good support for an application may have that application rejected and have no choice but to go through an expensive appeal, yet examples (again as in this book) show that properly managed engagement can generate widely supported and fully acceptable applications. Significant amounts of money can be saved through proper engagement but we are faced with a general societal problem that the costs of poor practice are never recorded because 'it is just what happens, isn't it?', so nobody ever thinks it worth accounting for.

But demonstrating that good engagement is value for money still does not answer the question: Where should the money (or resources, see later) come from? The fundamental problem for a public authority in assessing how to attribute resources is that they may *invest* but they do not secure the *benefits*; the public, NGOs, even landowners and developers do. If one could argue a cost of *not* doing the engagement then the direct and indirect financial benefits of doing it could well be considerable. However, as above, that evidence is not available.

The situation for an applicant/developer is different. Time is money and much of that will be formally accounted. So, for example, having paid to buy a piece of land and then paying interest on a loan while waiting for income from the development, securing planning permission in eight weeks rather than 16 is of direct financial benefit; it is interest (money) saved. Gaining community support can enable a quicker start on site, less hassle during the construction period and even quicker returns. On one of this author's projects the intense local engagement led to some of the houses in a development being sold almost before they were built, so sales and marketing costs were reduced. It is also usual for an applicant/developer to pay most of the costs of pre-application engagement, so these do not bear on the public purse.

The proportional costs of engagement, i.e. what percentage of the costs of producing a plan or application might go on engagement, are as difficult to pin down as actual costs but, in general, they are very, very little, almost inconsequential in the overall picture. In the case of the developer quoted above where some houses were pre-sold, the engagement cost less than 0.5 per cent of the budget just for the application stage. Even for a major strategic plan the proportional costs are likely to be of the same minor order.

Just by way of example, the costs in fees for this author's small company for managing and delivering engagement on a strategic plan came close to £30,000. If, as has been shown, this can save an authority around £200,000 in legal costs for an examination, that is well worth it. For a complex development project the cost was just under £25,000. For a medium-sized development project the cost was close to £6,000. Once again, speedy and largely uncontested applications create savings, if less easy to cost.

In actually managing any engagement work, and hence in budgeting for it, some simple factors need to be borne in mind. Consultant day rates may be more than the equivalent day rates for employees, but carefully selected (specialist) consultants may well be able to do in three days what would take the (generalist) internal staff eight or ten days to do. However, the engagement designers and deliverers, whether internal or external, are not the only personnel whose time and costs must be considered. The plan-making team or the developer's project manager, architect, ecologist etc. will almost certainly not only need to produce particular information (maps, sketches etc.) for the engagement work but also probably attend workshops and drop-ins. All of that is far too often overlooked and should also be costed.

There are then various direct costs that need to be planned for and monitored: consultant or officer expenses (travel, accommodation and subsistence), venue hire, catering, exhibition screens, materials, bulk mailing, website setting up and management, insurances and so forth.[2]

There is, however, another side to all of this. Most engagement practice depends to a greater or lesser extent on completely voluntary time from those involved. It may be to a *lesser* degree for those whose jobs are linked to an issue because they may be paid to do that, for example, when a highways officer or a paid staff member from a NGO attends a workshop. In almost every situation, however, it is to a *greater* degree because almost all community input is from people giving their own free time. Here are two examples:

- An informal estimate of the community time that goes into producing what are called Parish Plans[3] suggests nearly 4,500 hours of voluntary time. Using figures that some organisations propose suggests this 'sweat equity' is the equivalent of £75,000 cash.[4]
- In another case, engaging several communities in housing site selection for a strategic plan generated an extraordinary 384 days worth of 'free' time.

Though figures such as this are not really the point, it is essential to recognise that, in many situations, the community and/or the stakeholders may well be contributing as much time as the professionals, perhaps more. This book is about focusing on the quality of relationships that are established in effective collaborative working, so genuine respect for the voluntary commitment that many (most?) people make is essential in shaping those relationships. This should affect all behaviours at all stages, from the manner of dealing with all involved to the value given to their input, even down to the common courtesy of providing refreshments for workshop participants.

Notes

1   Andrew Acland.
2   And for some high conflict cases for this author, on-site security staff!
3   These are wide-ranging community plans, not land use or spatial plans.
4   Caution is urged here because some members of the community sector argue that attributing monetary values to voluntary time is ethically inappropriate and should be avoided.

Appendix 4
Becoming a Practitioner

Ideally of course, by this stage the reader will be champing at the bit, asking 'where do I go for some training in engagement generally, or process design or facilitation?' Sadly, there seem to be few good answers to this question. The phrase is 'seem to be' because this author has a good but by no means exhaustive knowledge of what is available. So the advice that follows is partial and good networking, both nationally and locally, and especially via local professional networks, is essential.

Comments have already been made on the lack of skills training in this area in initial professional education. Hopefully, that will get better as engagement and collaborative working become more the norm. There again, initial professional education should never be just about skills; it must cover theory and principles or people will not know where, why, whether and how to use any skills. In addition, there are some skills that are best or can only be learned at early or mid-career stage and on the job, and that is true for much of what is in this book.

In general, there *are* courses available (if few in number or difficult to find) on **facilitation** skills. This does not need to be dedicated to planning specifically because such skills are generic. It is worth contacting InterAct Networks for advice as much as actual training (all contact details are listed below). Some are run, if only occasionally, by business schools at universities, for example, at the University of the West of England. Those run by the ICA come highly recommended and more tailored courses can be provided by Icarus and Rowena Harris/Cathy Williams. There are also many books offering guides to facilitation that are not listed here but are easily accessed through a web search, though all come with the usual warning about the limits of book (or even course) learning.

Caution is needed on courses about **consultation as a whole** because that is what almost all of them are, i.e. not about engagement or collaborative working but basic consultation. Once again the ICA is a good point of contact, as is the Consultation Institute. Beware, however, because some courses appear to promote just one particular version or model and this can be limiting, while others do little more than teach methods (just 'ingredients' again).

Given this last comment, perhaps the biggest gap is in terms of training for **overall process design**; this remains very much the exception. Useful contacts are again Icarus and Rowena Harris/Cathy Williams, but also try InterAct Networks.

InterAct Networks also appears to be the only organisation offering training (in the broadest sense) on **organisational capacity building for engagement**.

Courses targeted at planners do take place but in a rather ad hoc manner, perhaps put on by a higher education institute or a Royal Town Planning Institute region (or both working together). Some events are entitled training (and qualify as such under continuing professional education criteria) but are really more like conferences with a few speakers and perhaps a short (often very short) practical workshop session. Sadly, this author is not aware of any regular mid-career training courses on engagement targeted specifically at planners, architects, engineers and others.

Contacts (see endnotes for web links):

- InterAct Network[1]
- The Consultation Institute[2]
- ICA[3]
- UWE Business School[4]
- Icarus[5]
- Rowena Harris and Cathy Williams[6]

Notes

1. Go to: www.interactnetworks.co.uk
2. Go to: www.consultationinstitute.org
3. Go to: http://www.ica-uk.org.uk/facilitation-training/
4. Go to: http://courses.uwe.ac.uk/Z20000048
5. Contact Steve Smith at steve@icarus.uk.net
6. Contact Rowena Harris at rowenaharris@btconnect.com or Cathy Williams at cathy@indras-net.co.uk

Appendix 5
Further Reading

References have been included throughout this book. Many are there to provide evidence and further information about specific points. However, some are generally valuable and refer to books that could be regarded (and are regarded by this author) as good overall sources of advice and ideas. Those are, therefore, repeated below along with others that provide further and sometimes different ideas and suggestions, ways of working, issues to consider and so forth; they are recommended reading. In most cases there is a short note about the book to help direct the reader to those that may prove most useful.

Acland, A. (1990) *A Sudden Outbreak of Commonsense*. London: Hutchinson.

This is a short, very direct and highly readable outline of why adversarial approaches are so damaging and expensive (in all senses) and why consensus building processes can bring enormous benefits. The author has been involved in helping to resolve numerous high-conflict situations, few of them within the planning world but the key points undoubtedly also apply to land use planning. As mentioned in Appendix 1, it is one of this author's 'top four' books on this topic.

Chambers, R. (2002) *Participatory Workshops*. London: Earthscan.

This is an extremely practical book based on wide experience on a variety of issues. Its key points are entirely consistent with those covered in this book. Chambers covers considerably more detail about workshop preparation, management and facilitation than possible here.

Creighton, J. (2005) *The Public Participation Handbook*. San Francisco: Jossey-Bass.

This is one of the 'gold standard' books by a highly experienced and very well practiced author. It is rooted in general in USA contexts but can be read and be useful for other contexts as well. Given its focus on participation rather than collaborative working, it covers more ground and provides more detail than this book, for example, about methods at all stages.

Driskell, D. (2002) *Creating Better Cities with Children and Youth*. London: Earthscan.

Though clearly targeted at work with children and young people, this book provides excellent coverage of principles and practice for engagement work with any age group. In fact, one might say that approaches that successfully engage young people ought to be seen as a bottom line for general practice.

Forester, J. (1999) *The Deliberative Practitioner*. Cambridge, Massachusetts: MIT Press.
Forester, J. (2009) *Dealing with Difference: Dramas of mediating public disputes*. Oxford: Oxford University Press.

John Forester has been mentioned in an earlier Appendix, citing the first of the two books above as one of this author's 'top four'. His focus is very much on land use planning and both books include some highly detailed and perceptive examples, using stories to fully explain key points. The books do not elaborate processes and methods as such but much can be drawn from them to inform practical approaches.

Healey, P. (1997) *Collaborative Planning: Shaping places in fragmented societies*. London: Macmillan.
Healey, P. (2010) *Making Better Places: The planning project in the twenty-first century*. Basingstoke: Palgrave Macmillan.

Once again, the first of these has been cited earlier as one of this author's 'top four'. Both are firmly theoretical and challenging to read but they provide a solid underpinning on the why and how of collaborative working. The second is more about substantive planning issues rather than processes but it is still underpinned by the principles of collaborative working.

Innes, J. and Booher, D. (2010) *Planning with Complexity: An introduction to collaborative rationality for public policy*. London: Routledge.

This is the final one of this author's 'top four'. As with Healey's books, it is firmly theoretical although it also includes numerous practical examples. By focusing on public policy generally, and mainly in a USA context, rather than (and despite the title) planning as in land use planning it is a very valuable addition to fundamental thinking about collaborative working.

Kaner, S. (2014) *The Facilitator's Guide to Participatory Decision Making*. San Francisco: Jossey-Bass Business & Management Series.

This is as good as it gets in terms of detailed guidance for anybody wishing to get involved seriously in facilitation. As well as outlining (in detail) many of the key skills, Kaner also places this in the context of a whole process journey from start to finish (through the 'groan zone').

Sarkissian, W. and Hurford, D. (2010) *Creative Community Planning: Transformative engagement methods for working at the edge*, London: Earthscan.

This is included in the list because, although it deals only with methods, the authors offer a whole repertoire of creative ideas (and the rationale behind them) that moves well beyond what has been covered in this book.

Susskind, L., McKearnan, S. and Thomas-Larmer, J. (Eds.) (1999) *The Consensus Building Handbook*. London: Sage.

For many, this is the absolute key book to this topic. It is very long and very detailed (and expensive) but includes just about everything from principles through practical details to a variety of long and thorough case studies, some but not all in the general territory of planning.

Wates, N. (2008) *The Community Planning Event Manual*. London: Earthscan.
Wates, N. (2014) *The Community Planning Handbook*. London: Earthscan.

Nick Wates' work, his two books listed here and the website,[1] provide the key complement to this book. They offer a superb compendium of methods (broad and narrow), ways to combine them and links on to other books, websites and people. Reading this book and Wates' books together provides full coverage from core questions of 'why' right down to the need for tape and scissors when running a workshop!

The following useful books are also valuable:

Brodie, E., Cowling, E. and Nissen, N. (2009) *Understanding Participation: A literature review*. London: Involve.[2]
Carley, M. and Bayley, R. (2009) *Urban Extensions, Planning and Participation*. Joseph Rowntree Foundation.[3]
Great Britain, Department of Communities and Local Government (2006) *Achieving Successful Participation: A literature review*. London: Department of Communities and Local Government.
New Economics Foundation (1998) *Participation Works: 21 techniques of community participation*. London: New Economics Foundation.
Pearce, B. (2010) 'Cost-effective Community Involvement in Planning', *Town and Country Planning Association Journal*, 76 (10), pp. 345–349.
Ricketts, S. and Field, D. (2012) *Localism and Planning*. London: Bloomsbury Professional.
Willow, C. (2002) *Participation in Practice*. London: The Children's Society.

Notes

1. Go to: www.communityplanning.net
2. Go to: www.involve.org.uk
3. Go to: www.jrf.org.uk

Index

Arbitrator, 124, 125
Arrivals exercises, 127, 135, 136, 147
Awareness, awareness-raising, 4, 33, 65, 71, 75, 80, 82, 98, 168, 171, 177, 180, 181, 184, 187, 201, 204-209

Barriers, 10, 30, 34, 40, 63, 103, 133, 176, 182, 186, 219
Benefits, 10, 14, 16
Brainstorming, 102-104, 106, 107, 120, 139
Briefing and briefs, 18, 46, 97-99, 101-103, 106-108, 113, 120, 123, 137-141, 147, 149, 163
Building blocks (of a process), 69-71, 79, 80, 133, 149, 164, 184, 197

Capacity building and capacity release, 2, 4, 7, 11, 14, 47, 69, 173-177, 179, 180-182, 186-188, 190, 196-198, 201-203, 208, 232
Carousel, 107, 108
Chairperson, 124, 125
Clients, 31, 43, 44, 49, 51, 53, 100, 153, 155, 156, 166, 171
Commissioners, 16, 17, 43, 44, 47, 51, 53, 153, 155, 173, 223
Commitment, 7, 17, 19, 39, 40, 44, 45, 54, 68, 137, 175, 177-181, 183, 186, 187, 201, 224, 230
Common ground, 17, 19, 39, 132, 134, 142, 150
Communications, 48, 49, 59, 130, 165, 186, 187, 214, 225, 226
Consensus, consensus building, 1, 10, 11, 13, 15, 17-19, 21, 22, 36, 39, 66, 83, 85, 115, 117, 129, 212, 213, 215, 221, 228, 233, 235
Consultees, 5, 42, 52, 54, 56, 57, 60, 61, 66, 70, 72, 75, 80, 161, 168, 203, 204, 208, 218, 219-223, 225, 227
Cross-cutting, 108

Database, 60, 66, 80, 94, 189, 194, 196, 197, 203, 205, 208
Decide-Announce-Defend, 21, 22, 28, 34-37, 40
Deliberation, 1, 10, 40, 48, 77, 215
Developers, 28, 32, 43, 173, 176, 178, 193, 206, 208, 210, 228
Dialogue, 1, 4, 10, 15, 25, 34, 40, 48, 73, 77, 78, 85, 86, 88, 121, 156, 195, 212, 214
Difficult people, 146, 147
Drop-in, 74, 81, 88, 93-96, 99, 117, 119, 126, 155, 158, 160, 168, 170, 205, 206, 209, 210, 229

Elected members or representatives and councillors, 4, 17, 29, 32, 33, 37, 38, 40, 54, 56, 57, 61, 64, 65, 72, 153, 161, 171, 174, 176, 178, 182, 183, 191, 196, 200, 201, 202, 204, 205, 207, 213, 224, 229,
Engage-Deliberate-Decide, 22, 28, 34-36, 38, 40
Evaluation, 1, 6, 45, 47, 65, 89, 95, 152-157, 160, 162, 165, 170, 171, 172, 186, 202, 203, 207
Evidence, 9, 16, 18, 27, 30, 31, 34, 54, 70, 124, 153, 161, 176, 184, 203, 228, 233

Examples, 4, 5, 7, 10, 14, 26, 30-32, 34-36, 40, 41, 58, 93, 97, 100, 112, 132, 134, 143, 156, 165, 168, 171, 175, 187-190, 195, 199, 201, 210, 213, 221, 222, 228, 230, 234

Facilitator networks, 188, 198
Feedback, 11, 18, 19, 34, 39, 55, 60, 69, 72, 93, 98, 100, 102, 107, 120, 124, 140, 141, 150, 152-155, 159, 160, 168, 191, 192, 196, 201, 218, 219
Funding, 16, 35, 42, 46, 47, 49, 54, 66, 103, 207

Groan zone, 83-86, 109, 235
Ground rules, 102, 103, 106, 107, 136, 147, 149

Hard to reach or engage (groups), 26, 62, 75, 78, 81, 206

Inclusiveness, 17, 19, 23, 25
Independence, 16, 43-46, 83, 130, 149, 149, 154, 162, 185, 188, 212

Legal, (or quasi-legal) context, 2, 4-7, 10, 26, 27, 30, 39, 40, 57, 60, 152, 154, 161,164, 172, 190, 203, 217-221, 225, 226, 229
Levels (of engagement), 5, 10-12, 14, 15, 23, 39, 75, 224
Listening, 15, 17, 51, 117, 130-132, 150, 216

Managing engagement, involvement or process, 3, 42, 44, 52, 67, 69, 163, 171, 220, 229
Managing people or group/group work, 6, 138, 140
Media, communications media, social media, 32, 34, 40, 48, 49, 56, 62, 63, 66, 71, 74, 95, 117, 160, 208
Mediator, 124, 125, 215
Mind mapping, 104

Moveable notes/Post-Its, 72, 76, 103, 127, 209

NGOs and voluntary sector, 8, 33, 43, 59, 60, 69, 72, 120, 175, 176, 188, 195, 203, 218, 228

Options, 6, 17, 25, 28, 38, 39, 69, 73, 82, 83, 85, 86, 93, 95, 99, 139, 144, 156, 171, 184, 220, 221
Organisational capacity building and learning, 153, 175, 176, 179, 182, 186, 187, 198, 232
Organisational change, 2, 186
Organisational culture, 13, 180
Outcomes, 3, 4, 10, 14-19, 2931, 34, 39, 40, 43, 50, 64-66, 69, 71, 72, 76, 86, 89, 91, 97, 98, 101, 106, 160-162, 165, 167, 168, 192, 199, 205, 209
Outreach, 75, 81

Participation and participate, 1, 8, 10, 11, 12, 26, 27, 30, 40, 41, 44, 45, 63, 66, 71, 172, 179, 190, 198, 227, 233, 235
Presenting problem, 42, 49, 50, 51, 65
Principles, 4, 5, 7, 10, 16, 19, 23, 25, 34, 38-40, 43, 44, 47, 53, 65, 68, 69, 72, 117, 130, 153-155, 162, 163, 171, 173, 174, 176, 182, 187-192, 194, 195, 218, 226, 231, 234, 235
Priority-setting, ranking and weighting, 95, 109, 111, 191, 205
Private sector, 3, 54, 59, 60, 176, 178, 186, 203
Process design and designers, designing processes, 4, 19, 42, 44, 49, 55, 67-71, 79, 86, 149, 153, 155, 162, 216, 167, 180, 187, 191, 194, 203, 204, 208, 216, 231, 232
Professionalism, 14, 31, 176
Public or local authorities, 3, 28, 35, 36, 37, 59, 62, 69, 129, 173, 176,

178, 186, 187, 189, 191, 193, 199-203, 221-224
Public relations and PR, 15, 48, 49, 70, 187

Questionnaire(s), 15, 18, 24, 26, 29, 51, 58, 74, 85, 86, 155, 156, 159, 160, 168, 210
Questioning, 5, 17, 51, 130, 132, 150
Questions to ask, 5, 52, 64, 160

Reach/reaching agreement or decisions, 35, 38, 100, 117, 143, 213
Recording, 97, 98, 132, 139, 140, 142, 143
Reframing, 130, 133, 134, 140, 150
Reports, reporting, 1, 6, 18, 19, 26, 28, 30, 34, 39, 48, 69, 70, 80, 96, 137, 144-146, 148, 152, 156, 160, 161, 164-168, 172, 206, 226
Resources, resourcing, 5, 7, 14, 16, 28, 40, 46, 53-55, 69, 73, 100, 129, 173, 174, 183, 186, 191, 194, 200, 203, 218, 228
Rhubarb sack, 142, 143

Scope, 5, 7, 12, 16, 17, 19, 20, 38, 39, 42, 52-55, 57, 61, 67, 72, 85, 69, 101, 106, 121, 123, 143, 189, 192, 194, 212
Stakeholder analysis, 75, 77
Stakeholder(s), 4, 5, 7, 11, 12, 15, 16, 24, 25, 26, 36, 38, 39, 42, 44, 45, 49, 54, 56-59, 61, 64, 66, 71, 72, 73, 75, 76, 77, 78, 81, 82, 84-86, 88, 90, 93, 95, 115, 116, 117, 121, 124, 126, 130, 153, 155, 162, 163, 167, 168, 173, 174, 177-179, 181, 189, 193, 203, 205, 208-210, 215, 218, 230
Statement of Community Involvement or SCI, 27-29, 39, 44, 60, 62, 164, 166, 171, 190, 195, 207, 223
Steering Group, 36, 44, 49, 59, 72, 73, 76, 80, 83, 84, 154, 162, 163, 167, 197, 204
Strategy grids, 111, 116, 138
SWOT, 106-109

Timetable and timetabling, 14, 55, 69, 89, 90, 96, 167, 178, 191
Tracking, 84, 142

Venue(s), 18, 46, 53, 54, 56, 89, 90, 123, 126, 127, 150, 185, 193, 205, 229
Visioning, 64, 101, 109

Wall-sheets or sheets, 93, 94, 96, 103, 108, 110, 112, 113, 117, 119, 127, 128, 131, 140, 142, 145, 157, 163
Waterton, 7, 39, 55, 56, 61, 63, 64, 70, 72, 73, 75, 76, 79, 80, 81, 83, 92, 93, 96, 98-100, 119, 123, 145, 163, 171, 187, 195
Websites, Facebook, Twitter, 15, 48, 49, 56, 67, 85, 89, 94, 95, 96, 159, 160, 165, 167, 168, 194, 196, 205, 206, 209, 210, 217, 229, 235
Work in-breadth, 74, 75, 99
Work in-depth, 26, 40, 73, 99